SpringerBriefs in Mathematical Physics

Volume 43

SpringerBriefs are characterized in general by their size (50–125 pages) and fast production time (2–3 months compared to 6 months for a monograph).

Briefs are available in print but are intended as a primarily electronic publication to be included in Springer's e-book package.

Typical works might include:

- An extended survey of a field
- A link between new research papers published in journal articles
- A presentation of core concepts that doctoral students must understand in order to make independent contributions
- Lecture notes making a specialist topic accessible for non-specialist readers.

SpringerBriefs in Mathematical Physics showcase, in a compact format, topics of current relevance in the field of mathematical physics. Published titles will encompass all areas of theoretical and mathematical physics. This series is intended for mathematicians, physicists, and other scientists, as well as doctoral students in related areas.

Editorial Board

More information about this series at https://link.springer.com/bookseries/11953

Yukinobu Toda

Recent Progress on the Donaldson–Thomas Theory

Wall-Crossing and Refined Invariants

 Springer

Yukinobu Toda
Kavli IPMU
University of Tokyo
Chiba, Kashiwa-shi, Japan

ISSN 2197-1757   ISSN 2197-1765   (electronic)
SpringerBriefs in Mathematical Physics
ISBN 978-981-16-7837-0   ISBN 978-981-16-7838-7   (eBook)
https://doi.org/10.1007/978-981-16-7838-7

This Springer imprint is published by the registered company Springer Nature Singapore Pte Ltd.
The registered company address is: 152 Beach Road, #21-01/04 Gateway East, Singapore 189721,
Singapore

Preface

This monograph is an expository text on Donaldson–Thomas (DT) theory on Calabi–Yau (CY) 3-folds. We focus on curve counting DT invariants on CY 3-folds, wall-crossing phenomena in the derived category and cohomological refinements.

The DT invariants virtually count stable coherent sheaves on CY 3-folds, which were introduced by Thomas [163] around 1998. They give a complex analytic version of Casson invariants for real 3-manifolds, and a higher-dimensional analogue of Donaldson invariants for real 4-manifolds. The theory was extended to counting semistable sheaves by Joyce–Song [92]. Around 2003, Maulik–Nekrasov–Okounkov–Pandharipande (MNOP) [126] proposed a conjecture relating rank one DT invariants and Gromov–Witten (GW) invariants; both count curves on CY 3-folds. Since then, DT theory has drawn much attention in algebraic geometry and mathematical physics.

On the other hand, around 2002, Bridgeland [27] introduced the notion of stability conditions on triangulated categories, especially derived categories of coherent sheaves on CY 3-folds. The motivation of stability conditions comes from mirror symmetry. The set of stability conditions on a CY 3-fold forms a complex manifold, which is expected to be related to the stringy Kähler moduli space. Bridgeland stability conditions also turned out to be important in DT theory. A stability condition defines the notion of (semi)stable objects in the derived category, and one can try to count them. In this way, one can try to extend DT theory from sheaves to complexes of sheaves. For example, the Pandharipande–Thomas (PT) invariant [146] is one of the invariants that counts certain two-term complexes and plays an important role in the MNOP conjecture.

By extending DT theory from sheaves to complexes, we have more flexibility in the theory. The DT invariants counting Bridgeland (semi)stable objects depend on a choice of a stability condition, and it is a natural problem to investigate such dependency. In general, there is a wall–chamber structure on the space of stability conditions such that the invariants are constant if a stability condition lies in a chamber but may change when it crosses a wall. The wall-crossing formula, established by Kontsevich–Soibelman [108] and Joyce–Song [92], turned out to imply several important properties of the generating series of curve counting DT theory, e.g. DT/PT correspondence, rationality of PT generating series, flop formula, etc.

The original DT invariants were defined via integrals of zero-dimensional virtual fundamental classes on the moduli spaces of stable sheaves on CY 3-folds. Later, Behrend [11] showed that the DT invariants are also defined via the weighted Euler characteristics of certain constructible functions (called Behrend functions) on the above moduli spaces. Now, these moduli spaces are known to be locally written as critical loci of some functions, and Behrend functions coincide with point-wise Euler characteristics of the perverse sheaves of vanishing cycles. Based on this fact, Kontsevich–Soibelman [109] proposed a refinement of DT invariants, called cohomological DT invariants. They are defined by gluing the locally defined perverse sheaves of vanishing cycles through a choice of orientation data. The foundation of such a cohomological refinement was established in [24]. Such a refinement was used in [128] to give a mathematical definition of Gopakumar–Vafa invariants, which conjecturally control both GW invariants and PT invariants.

In this monograph, we review the above developments on DT invariants. The target readers are Ph.D. students who are interested in working with DT theory, researchers of algebraic geometry working in areas other than DT theory but who want to quickly learn the subject and physicists who are interested in the current mathematical development of DT theory. We do not give any detail on the proofs of mathematical statements, but rather give relevant references so that readers interested in details can access the proofs. We remark that, due to the page limitation, we could not include all the topics of DT theory, and for example, this monograph does not contain the topics of the degeneration formula of DT invariants [119], DT theory for CY 4-folds [23, 42, 139] and the relation with cluster algebras [134]. There also have been several survey articles on DT theory so far, e.g. [130, 140, 147, 159, 172], which are also recommended to readers.

I would like to thank Tom Bridgeland, Dominic Joyce, Davesh Maulik, Rahul Pandharipande and Richard Thomas for discussions on the subject of this monograph for more than a decade. I would like to also thank Yalong Cao and Hsueh-Yung Lin for several comments on the preliminary version of the manuscript, and anonymous referees for carefully checking the manuscript and giving valuable comments. The author is supported by the World Premier International Research Center Initiative (WPI initiative), MEXT, Japan, and a Grant-in-Aid for Scientific Research grant (No. 19H01779) from MEXT, Japan.

Kashiwa, Japan   Yukinobu Toda

Contents

Chapter 1
Donaldson–Thomas Invariants on Calabi–Yau 3-Folds

Abstract The Donaldson–Thomas invariants are introduced by Thomas [163] as a holomorphic analogue of Casson invariants [162]. They count stable coherent sheaves on CY 3-folds. More precisely, they are defined as integrations of zero-dimensional virtual fundamental classes on moduli spaces of stable coherent sheaves. In the rank one case, the generating series of DT invariants are conjectured to be related to the generating series of Gromov–Witten invariants, which is proved in many cases. In this chapter, we give an overview of DT invariants on CY 3-folds and GW/DT correspondence conjecture.

1.1 Moduli Spaces of Semistable Sheaves on Algebraic Varieties

In this section, we review the theory of (semi)stable sheaves on algebraic varieties. The main reference of this section is Huybrechts–Lehn's book [82]. We are also free to use the terminology of stacks, whose basic references are [112, 141]. Throughout this monograph, all the schemes or stacks are defined over $\mathbb{C}$.

Let X be a smooth projective variety, $\mathcal{O}_X(1)$ an ample line bundle on it and set $H := c_1(\mathcal{O}_X(1))$. For $E \in \mathrm{Coh}(X)$, by the Riemann–Roch theorem its *Hilbert polynomial* is written as

$$
\chi(E(m)) = \int_X \mathrm{ch}(E(m))\mathrm{td}_X
$$
$$
= a_d m^d + a_{d-1} m^{d-1} + \cdots + a_0,
$$

for some $a_0, \ldots, a_d \in \mathbb{Q}$. Here d is the dimension of the support of E and $a_d \neq 0$. The *reduced Hilbert polynomial* is defined by

$$
\overline{\chi}(E(m)) := \chi(E(m))/a_d \in \mathbb{Q}[m].
$$

For polynomials $f_i \in \mathbb{Q}[m]$ for $i = 1, 2$, we write $f_1 \prec f_2$ if either

© The Author(s), under exclusive license to Springer Nature Singapore Pte Ltd. 2021
Y. Toda, *Recent Progress on the Donaldson–Thomas Theory*,
SpringerBriefs in Mathematical Physics 43,
https://doi.org/10.1007/978-981-16-7838-7_1

$$\deg(f_1) > \deg(f_2), \text{ or } \deg(f_1) = \deg(f_2), \ f_1(m) < f_2(m), \ m \gg 0.$$

Definition 1.1 An object $E \in \mathrm{Coh}(X)$ is called:

(i) *H-(semi)stable* if for any non-zero subsheaf $E' \subsetneq E$ we have the inequality $\overline{\chi}(E'(m)) \prec (\preceq)\overline{\chi}(E(m))$,
(ii) *strictly H-semistable* if it is H-semistable but not H-stable,
(iii) *H-polystable* if it is H-semistable and a direct sum of H-stable sheaves.

Remark 1.1 If $E \in \mathrm{Coh}(X)$ is H-semistable, it must be *pure*, i.e. for any non-zero subsheaf $E' \subset E$ we have dim $\mathrm{Supp}(E') = \dim \mathrm{Supp}(E)$. Indeed, if dim $\mathrm{Supp}(E') < $ dim $\mathrm{Supp}(E)$, we have $\overline{\chi}(E(m)) \prec \overline{\chi}(E'(m))$ so E' destabilizes E.

Remark 1.2 Suppose that dim $X = 1$ and E is a torsion-free sheaf on X. Then $\overline{\chi}(E(m)) = m + \mu(E) + (const)$ where $\mu(E) = \deg(E)/\mathrm{rank}(E)$. Therefore E is H-(semi)stable if and only if for any non-zero $E' \subsetneq E$, we have $\mu(E') < (\leq)\mu(E)$.

An important property of (semi)stable sheaves is that they form the building blocks for all the coherent sheaves on X. For an object $E \in \mathrm{Coh}(X)$, let us consider a filtration

$$0 = E_0 \subset E_1 \subset \cdots \subset E_n = E \tag{1.1}$$

with subquotient $F_i = E_i/E_{i-1}$.

Definition 1.2 A filtration (1.1) is called:

(i) *Harder–Narasimhan (HN) filtration* if each F_i is H-semistable with $\overline{\chi}(F_i, m) \succ \overline{\chi}(F_{i+1}, m)$ for all i.
(ii) *Jordan–Hölder (JH) filtration* if F_i is H-stable with $\overline{\chi}(F_i, m) = \overline{\chi}(E, m)$.

For any $E \in \mathrm{Coh}(X)$, there exists an HN filtration unique up to isomorphism (see [82, Theorem 1.3.4]]). Moreover, for any H-semistable sheaf E, there exists a JH filtration which is not unique but the direct sum of F_i is unique up to isomorphism (see [82, Proposition 1.5.2])

In what follows, we denote by Γ the image of the Chern character map

$$\Gamma := \mathrm{Im}(\mathrm{ch}\colon K(X) \longrightarrow H^{2*}(X, \mathbb{Q})). \tag{1.2}$$

For each $v \in \Gamma$, we denote by $\mathcal{M}_X(v)$ the 2-functor

$$\mathcal{M}_X(v)\colon (Sch/\mathbb{C}) \longrightarrow (Groupoid) \tag{1.3}$$

sending a $\mathbb{C}$-scheme T to the groupoid of T-flat families $\mathscr{E} \in \mathrm{Coh}(X \times T)$ such that any restriction $\mathscr{E}_t \in \mathrm{Coh}(X)$ by $X \times \{t\} \hookrightarrow X \times T$ has Chern character v. The morphisms in the above groupoid are given by isomorphisms $\mathcal{E} \xrightarrow{\cong} \mathcal{E}'$. It is well-known that the above 2-functor forms an Artin stack locally of finite type (see [112, Theorem 4.6.2.1]).

Remark 1.3 In this monograph, we will freely use the term *Artin stack* whose basic reference is [112]. For readers who are not familiar with stacks, it will be enough to consider an Artin stack as something like 'a variety with an action of an algebraic group' or some kind of gluing of such data. Indeed, a variety Y with an action of G determines a stack $[Y/G]$ called a *quotient stack*. Under some special situations, an Artin stack is locally a quotient stack (see [2]).

The Artin stack (1.3) is neither of finite type nor separated, so it is not a good geometric space we should consider. On the other hand, we have substacks

$$\mathcal{M}_X^{H\text{-st}}(v) \subset \mathcal{M}_X^{H\text{-ss}}(v) \subset \mathcal{M}_X(v) \tag{1.4}$$

consisting of $\mathcal{E} \in \mathrm{Coh}(X \times T)$ such that $\mathcal{E}_t$ is H-(semi)stable. Then the above substacks are open substacks of finite type (see [82, Proposition 2.3.1, Theorem 3.3.7]). Moreover, as we will see below, $\mathcal{M}_X^{H\text{-ss}}(v)$ admits a good moduli space. Here a general definition of good moduli spaces is given as follows:

Definition 1.3 ([1, Definition 4.1]) For an Artin stack $\mathcal{M}$, its *good moduli space* is an algebraic space M together with a quasi-compact morphism $\pi : \mathcal{M} \to M$ satisfying that:

(i) the push-forward $\pi_* : \mathrm{QCoh}(\mathcal{M}) \to \mathrm{QCoh}(M)$ is exact,
(ii) the induced morphism $\mathcal{O}_M \to \pi_* \mathcal{O}_\mathcal{M}$ is an isomorphism.

Example

Suppose that a reductive algebraic group G acts on an affine scheme $\mathrm{Spec} R$. Then the quotient morphism

$$[\mathrm{Spec} R/G] \longrightarrow \mathrm{Spec} R /\!/ G := \mathrm{Spec} R^G \tag{1.5}$$

is a good moduli space for $[\mathrm{Spec} R/G]$ (see [1, Theorem 13.2]).

By [1, Example 8.7], there is a projective scheme $M_X^{H\text{-ss}}(v)$ parametrizing H-polystable sheaves together with the Cartesian diagram

$$\begin{array}{ccc} \mathcal{M}_X^{H\text{-st}}(v) & \lhook\joinrel\longrightarrow & \mathcal{M}_X^{H\text{-ss}}(v) \\ \downarrow & \square & \downarrow{\scriptstyle \pi_X^H} \\ M_X^{H\text{-st}}(v) & \lhook\joinrel\longrightarrow & M_X^{H\text{-ss}}(v). \end{array} \tag{1.6}$$

Here $M_X^{H\text{-st}}(v) \subset M_X^{H\text{-ss}}(v)$ is an open subscheme corresponding to H-stable sheaves. The vertical arrows are good moduli space morphisms, sending an H-semistable sheaf to the associated graded sheaf of the JH filtration. Since $\mathrm{Aut}(E) = \mathbb{C}^*$ for any H-stable sheaf E, the left morphism in (1.6) is a $\mathbb{C}^*$-gerbe.

Definition 1.4 (i) We say that $M_X^{H\text{-st}}(v)$ is a *fine moduli space* if $\mathcal{M}_X^{H\text{-st}}(v)$ is a trivial $\mathbb{C}^*$-gerbe over $M_X^{H\text{-st}}(v)$, i.e.

$$\mathcal{M}_X^{H\text{-st}}(v) \cong M_X^{H\text{-st}}(v) \times B\mathbb{C}^*. \tag{1.7}$$

(ii) We say that $M_X^{H\text{-st}}(v)$ satisfies the *ss=st condition* if the equality $M_X^{H\text{-ss}}(v) = M_X^{H\text{-st}}(v)$ holds.

Remark 1.4 The condition (1.7) is equivalent to saying that there is a universal sheaf on $X \times M_X^{H\text{-st}}(v)$. A universal sheaf is not unique, as it depends on a choice of a trivialization in (1.7), and any other universal sheaf differs by a tensor product with a line bundle on $M_X^{H\text{-st}}(v)$. The condition (1.7) is satisfied if the following condition holds (see [82, Theorem 4.6.5]):

$$\text{g.c.d.}\{\chi(E \otimes F) : [E] \in M_X^{H\text{-st}}(v), F \in K(X)\} = 1. \tag{1.8}$$

Example

Let $v = (v_0, v_1, \ldots)$, here $v_i \in H^{2i}(X, \mathbb{Q})$ satisfies $v_0 = 1$. In this case, any H-stable sheaf E with $\mathrm{ch}(E) = v$ is of the form

$$E \cong \mathcal{L} \otimes I_Z \tag{1.9}$$

for a line bundle $\mathcal{L}$ on X and a closed subscheme $Z \subset X$ with $\mathrm{codim}\, Z \geq 2$.

Indeed, for each H-stable sheaf E on X with $\mathrm{ch}(E) = v$, we have the injection $E \hookrightarrow E^{\vee\vee}$ and $E^{\vee\vee}$ is a line bundle $\mathcal{L}$ as it is a rank one reflexive sheaf. Therefore E is of the form (1.9), and we have the isomorphism

$$\mathrm{Pic}_X(v_1) \times \mathrm{Hilb}_X(e^{-v_1}v) \xrightarrow{\cong} M_X^{H\text{-st}}(v). \tag{1.10}$$

Here $\mathrm{Pic}_X(v_1)$ is the *Picard scheme* parameterizing line bundles $\mathcal{L}$ on X with $c_1(\mathcal{L}) = v_1$ and $\mathrm{Hilb}_X(v)$ is the *Hilbert scheme* parameterizing closed subschemes $Z \subset X$ with $\mathrm{ch}(I_Z) = v$ (see [105, I.1]). In particular, $M_X^{H\text{-st}}(v)$ is independent of H if $v_0 = 1$. Moreover, the condition $v_0 = 1$ easily implies that there is no strictly semistable sheaf E with $\mathrm{ch}(E) = v$ and the condition (1.8) is satisfied, i.e. $M_X^{H\text{-st}}(v)$ is fine and satisfies the ss=st condition.

1.2 Deformation–Obstruction Theory

Here we review a basic deformation–obstruction theory of coherent sheaves. For details, see [82, 2.A.6].

Let $p \in M_X^{H\text{-st}}(v)$ be a closed point corresponding to a H-stable sheaf E on X. The Zariski tangent space of $M_X^{H\text{-st}}(v)$ at p is by definition identified with the isomorphism classes of objects $\mathscr{E} \in \mathrm{Coh}(X \times \mathrm{Spec}\mathbb{C}[t]/t^2)$ which is flat over $\mathrm{Spec}\mathbb{C}[t]/t^2$ and $\mathscr{E}|_{X \times \{0\}} \cong E$. Then we obtain the element $\xi \in \mathrm{Ext}^1(E, E)$ as an extension class of the exact sequence in X

$$0 \longrightarrow t \cdot E \longrightarrow q_*\mathscr{E} \longrightarrow E \longrightarrow 0.$$

Here $q \colon X \times \mathrm{Spec}\mathbb{C}[t]/t^2 \to X$ is the projection. In this way, we have the map

$$T_p M_X^{H\text{-st}}(v) \longrightarrow \mathrm{Ext}^1(E, E).$$

The above map can be easily proved to be an isomorphism, so $T_p M_X^{H\text{-st}}(v)$ is identified with $\mathrm{Ext}^1(E, E)$.

Let $\psi \colon R' \twoheadrightarrow R$ be a surjection of local Artinian $\mathbb{C}$-algebras. We say that ψ is a *small extension* if $\mathrm{Ker}(\psi) \cdot \mathfrak{m}' = 0$, where $\mathfrak{m}' \subset R'$ is the maximal ideal. In this case, $\mathrm{Ker}(\psi)$ is a vector space over $R'/\mathfrak{m}' = \mathbb{C}$. A deformation theory of coherent sheaves implies that, given an object $\mathscr{E} \in \mathrm{Coh}(X \times \mathrm{Spec}R)$ flat over R, there exists a class (called *obstruction class*)

$$o(\mathscr{E}) \in \mathrm{Ext}^2(E, E) \otimes_{\mathbb{C}} \mathrm{Ker}(\psi)$$

such that $o(\mathscr{E}) = 0$ if and only if $\mathscr{E}$ extends to $\mathscr{E}' \in \mathrm{Coh}(X \times \mathrm{Spec}R')$ flat over R'. Moreover, the set of possible extensions $\mathscr{E}'$ is a torsor over $\mathrm{Ext}^1(E, E) \otimes_{\mathbb{C}} \mathrm{Ker}(\psi)$.

The above deformation theory roughly means that $M_X^{H\text{-st}}(v)$ is locally cut out by the $\mathrm{ext}^2(E, E)$-equations from $\mathrm{Ext}^1(E, E)$. Using gauge theory (see [63, 183]), we have the following more precise statement: there exist analytic open neighborhoods $p \in V \subset M_X^{H\text{-st}}(v), 0 \in U \subset \mathrm{Ext}^1(E, E)$ and an analytic map (called *Kuranishi map*)

$$\kappa \colon U \longrightarrow \mathrm{Ext}^2(E, E) \tag{1.11}$$

such that $V \cong \kappa^{-1}(0)$ as complex analytic spaces.

Remark 1.5 It is also well-known that many deformation theories are governed by *differential graded Lie algebras (DGLA)* . In our case, the DGLA which governs deformation theory of E is $\mathbf{R}\mathrm{Hom}(E, E)[1]$. Using Dolbeaut model to compute $\mathbf{R}\mathrm{Hom}(E, E)[1]$, we can construct its minimal L_∞-structure on $\mathrm{Ext}^*(E, E)$, and the map κ is naturally constructed from the above L_∞-structure. See [63, 183] for details.

The existence of the above deformation–obstruction theory may be packaged into the notion of (perfect) obstruction theory [14]. For a scheme M, the *truncated cotangent complex* $\tau_{\geq -1}\mathbb{L}_M$ on it is given by, for each open subset $U \subset M$ together with a closed embedding $U \hookrightarrow A$ for smooth A with defining ideal $I \subset \mathcal{O}_A$,

$$\tau_{\geq -1}\mathbb{L}_M|_U = (I/I^2 \longrightarrow \Omega_A|_U),$$

where $\Omega_A|_U$ is located in degree zero. Below we denote by $D^b(M)$ the bounded derived category of coherent sheaves on M (see [76, 81] for basic references on derived categories of coherent sheaves).

Definition 1.5 ([14]) An *obstruction theory* on a scheme M is a morphism

$$\phi: \mathscr{E} \longrightarrow \tau_{\geq -1}\mathbb{L}_M \tag{1.12}$$

in $D^b(M)$ such that $\mathscr{H}^{>0}(\mathscr{E}) = 0$, $\mathscr{H}^0(\phi)$ is an isomorphism and $\mathscr{H}^{-1}(\phi)$ is surjective. It is called *perfect* if $\mathscr{E}$ is locally quasi-isomorphic to a two-term complex of vector bundles $(\mathscr{E}^{-1} \to \mathscr{E}^0)$.

Remark 1.6 If $\mathscr{E} \to \tau_{\geq -1}\mathbb{L}_M$ is a perfect obstruction theory, then we have $\mathscr{H}^0(\mathscr{E}^\vee|_x) = T_{M,x}$ for each point $x \in M$. Also for a small extension $\psi: R' \twoheadrightarrow R$ and a morphism $f: \operatorname{Spec} R \to M$, there is an obstruction class $o(f) \in \mathscr{H}^1(\mathscr{E}^\vee|_x) \otimes \operatorname{Ker}(\psi)$ such that $o(f) = 0$ if and only if f can be extended to $\operatorname{Spec} R' \to M$, and the set of possible extensions is a torsor over $\mathscr{H}^0(\mathscr{E}^\vee|_x) \otimes \operatorname{Ker}(\psi)$. See [14, Theorem 4.5].

Example

Let A be a smooth scheme, $V \to A$ a vector bundle on it and $s: A \to V$ be its section. Let I be the image of $s: V^\vee \to \mathcal{O}_A$ and $U \subset A$ the subscheme given by $\mathcal{O}_U = \mathcal{O}_A/I$. Then we have the following commutative diagram:

$$\begin{array}{ccc} V^\vee|_U & \longrightarrow & \Omega_A|_U \\ {\scriptstyle s|_U}\downarrow & & \downarrow{\scriptstyle \mathrm{id}} \\ I/I^2 & \longrightarrow & \Omega_A|_U. \end{array} \tag{1.13}$$

By setting $\mathscr{E} = (V^\vee|_U \to \Omega_A|_U)$, the above diagram gives a perfect obstruction theory on U.

Remark 1.7 Any perfect obstruction theory on a scheme is locally of the form (1.13). This seems to be a well-known fact to experts, but there does not seem to be a reference proving this fact rigorously. Here is a brief argument suggested by one of the referees.

Locally, we embed M into a smooth scheme A with defining ideal I. Since the cone of the morphism (1.12) is of the form $P[2]$ for a $P \in \mathrm{Coh}(M)$, the morphism $\Omega_A|_M \to \tau_{\geq -1}\mathbb{L}_M$ uniquely lifts to a morphism $\Omega_A|_M \to \mathscr{E}$ whose cone is of the form $V^\vee[1]$ for a vector bundle V on M. So we may assume that $\mathscr{E}$ is of the form $(V^\vee \to \Omega_A|_M)$. Moreover (as $\mathscr{E}^i$ are locally projective), we can represent the morphism (1.12) as a morphism of complexes from $(V^\vee \to \Omega_A|_M)$ to $(I/I^2 \to \Omega_A|_M)$. Then extend V to a vector bundle $\widetilde{V} \to A$ and use the surjection $V^\vee \to I/I^2$ and projectivity of $V^\vee$ to lift it to $\widetilde{V}^\vee \to I$. This defines a section of $\widetilde{V}$ cutting out $M \subset A$.

Remark 1.8 By the Kuranishi map (1.11) and the diagram (1.13), the moduli space $M_X^{H\text{-st}}(v)$ admits a perfect obstruction theory analytic locally at each point $[E] \in M_X^{H\text{-st}}(v)$, where the obstruction space is $\mathrm{Ext}^2(E, E)$. However, these perfect obstruction theories may not glue to give a global perfect obstruction theory. For the existence of a gluing, the ranks of perfect obstruction theories must be locally constant, i.e. $\mathrm{ext}^1(E, E) - \mathrm{ext}^2(E, E)$ must be a locally constant function for $[E] \in M_X^{H\text{-st}}(v)$.

For example, this is achieved when $\dim X \leq 2$, as $\mathrm{Ext}^{\geq 3}(E, E) = 0$ in this case. Indeed, in this case, there is a natural perfect obstruction theory on $M_X^{H\text{-st}}(v)$ as in the same construction of Theorem 1.2 (see [132, Chap. 5]).

In the situation of the diagram (1.13), we have the following *normal cone*:

$$C_{U/A} = \mathrm{Spec}_U\left(\bigoplus_{k \geq 0} I^k/I^{k+1}\right) \longrightarrow U.$$

The normal cone is an affine cone over U which is of pure dimension $\dim A$, and a closed subcone in $V|_U$. Therefore we have the class $[C_{U/A}] \in A_{\dim A}(V|_U)$, where $A_r(-)$ is the *Chow group* of r-dimensional algebraic cycles modulo rational equivalence (see [64, Sect. 1.3]). The *localized Euler class* is defined by

$$e(V, s) := 0^![C_{U/A}] \in A_{\dim A - \mathrm{rank}(V)}(U),$$

where 0 is the zero section of $V|_U \to U$ and $0^!$ is the *Gysin pull-back* (see [64, Definition 3.3]). Its push-forward along with $U \hookrightarrow A$ is the usual Euler class of $V \to A$ (see [50, Lemma 7.1.5]). The virtual fundamental class is constructed as a gluing of these localized Euler classes:

Theorem 1.1 ([14, Sect. 5]) *Given a perfect obstruction theory* $\phi\colon \mathscr{E} \to \tau_{\geq -1}\mathbb{L}_M$, *there is a virtual fundamental class*

$$[M]^{\mathrm{vir}} \in A_{\mathrm{vd}}(M)$$

which is independent of a presentation of $\mathscr{E}$ *and* ϕ. *Here* $\mathrm{vd} := \mathrm{rank}(\mathscr{E})$ *is the virtual dimension.*

We will not discuss the construction of the virtual class, and refer to [14, Sect. 5] for details.

Remark 1.9 The virtual cycle can be also constructed from derived algebraic geometry, which was the original proposal by Kontsevich [107, Sect. 1.4.2]. Now derived algebraic geometry is a huge subject, and we just refer to the references [65, 66, 189].

For a derived scheme $\mathfrak{M}$, we have the classical truncation $M = t_0(\mathfrak{M})$ which is the usual scheme, and for each point $x \in M$ we have the *cotangent complex* $\mathbb{L}_{\mathfrak{M}}|_x$. A derived scheme $\mathfrak{M}$ is *quasi-smooth* if $\mathbb{L}_{\mathfrak{M}}|_x$ is concentrated on $[-1, 0]$ at each point $x \in M$. By [25, Theorem 4.1], $\mathfrak{M}$ is quasi-smooth if and only if it is Zariski locally of the form

$$\mathfrak{U} = \operatorname{Spec}\left(\cdots \longrightarrow \wedge^2 V^\vee \xrightarrow{s} V^\vee \xrightarrow{s} \mathcal{O}_A \right), \qquad (1.14)$$

where A is a smooth scheme and $V \to A$ a vector bundle with a section s, i.e. $\mathfrak{U}$ is the derived zero locus of s whose classical truncation is $U = (s = 0) \subset A$. The closed immersion $M \hookrightarrow \mathfrak{M}$ induces the morphism

$$\mathbb{L}_{\mathfrak{M}}|_M \longrightarrow \tau_{\geq -1} \mathbb{L}_M, \qquad (1.15)$$

which is locally of the form (1.13), so it is a perfect obstruction theory.

For a quasi-smooth derived scheme $\mathfrak{M}$, its *virtual structure sheaf* $\mathcal{O}_M^{\mathrm{vir}}$ is defined by

$$\mathcal{O}_M^{\mathrm{vir}} := \sum_{i \in \mathbb{Z}} (-1)^i \mathcal{H}^i(\mathcal{O}_{\mathfrak{M}}) \in K_0(M), \qquad (1.16)$$

which is well-defined since $\mathcal{O}_{\mathfrak{M}}$ is a cohomologically bounded from (1.14) and each cohomology is an $\mathcal{O}_M$-module. Here $K_0(M)$ is the Grothendieck group of coherent sheaves on M. It is proved in [49, Theorem 3.3] that (1.16) lies in the subgroup $K_0(M)_{\leq \mathrm{vd}}$ generated by sheaves whose supports have dimensions less than or equal to the virtual dimension of $\mathfrak{M}$, and the virtual fundamental cycle $[M]^{\mathrm{vir}}$ for the perfect obstruction theory (1.15) coincides with $\mathrm{cl}(\mathcal{O}_M^{\mathrm{vir}})$ for the cycle map $\mathrm{cl}(-)$.

Remark 1.10 By [188], the moduli space $M = M_X^{H\text{-st}}(v)$ admits a natural enhancement of a derived moduli scheme. There is also a different kind of enhancement of the moduli space M to a non-commutative direction, i.e. some non-commutative ringed space M^{nc} whose abelization gives M. We refer to [99, 111, 155] for formal non-commutative deformations of sheaves, [94, 151, 182] for global non-commutative moduli spaces of sheaves. There also exist applications to birational geometry and enumerative geometry [57, 79, 178, 184].

1.3 Donaldson–Thomas Invariants

In what follows, we assume that X is a smooth projective Calabi–Yau (CY) 3-fold. In this monograph, this means the following:

$$\dim_{\mathbb{C}} X = 3, \ K_X \cong \mathcal{O}_X, \ H^1(\mathcal{O}_X) = 0.$$

Here $K_X := \wedge^3 \Omega_X$ is the canonical line bundle on X. A typical example is a smooth hypersurface $X \subset \mathbb{P}^4$ of degree 5.

Remark 1.11 In the following the condition $H^1(\mathcal{O}_X) = 0$ is not essential, as otherwise the arguments apply for moduli spaces of determinant fixed stable sheaves. However, in order to simplify the notation and arguments, we keep the condition $H^1(\mathcal{O}_X) = 0$ in this monograph.

If X is a CY 3-fold, then by Serre duality we have the following isomorphism:

$$\mathrm{Ext}^1(E, E) \cong \mathrm{Ext}^2(E, E \otimes K_X)^{\vee} \cong \mathrm{Ext}^2(E, E)^{\vee} \qquad (1.17)$$

for any coherent sheaf E on X. Therefore $\mathrm{ext}^1(E, E) - \mathrm{ext}^2(E, E) = 0$ and from Remark 1.8, one may expect the existence of a global perfect obstruction theory on the moduli space $M_X^{H\text{-st}}(v)$. Moreover, the isomorphism (1.17) implies that the tangent and the obstruction spaces of $M_X^{H\text{-st}}(v)$ are dual to each other. Such a perfect obstruction theory should satisfy the following symmetric property:

Definition 1.6 A perfect obstruction theory $\mathscr{E} \to \tau_{\geq -1} \mathbb{L}_M$ on a $\mathbb{C}$-scheme M is called *symmetric* if there exists an isomorphism in $D^b(M)$

$$\theta : \mathscr{E} \xrightarrow{\cong} \mathscr{E}^{\vee}[1]$$

satisfying $\theta^{\vee} = \theta[1]$.

Example

In the setting of the diagram (1.13), we set $V = \Omega_A$ and $s = df$ for a regular function $f : A \to \mathbb{C}$. Then the vanishing locus $U \subset A$ of s is the critical locus of f. The resulting perfect obstruction theory

$$(T_A|_U \xrightarrow{\mathrm{Hess}(f)|_U} \Omega_A|_U) \longrightarrow \tau_{\geq -1} \mathbb{L}_U$$

is a symmetric perfect obstruction theory. Here $\mathrm{Hess}(f)$ is the Hessian of f.

Remark 1.12 The existence of a symmetric perfect obstruction theory on a $\mathbb{C}$-scheme M imposes a strong constraint on the singularities of M. It is proved in [15, Corollary 1.21] that if M is reduced, has only complete intersection singularities and admits a symmetric perfect obstruction theory, then M is smooth. In particular, by the above example, any reduced complete intersection singularity (which is not smooth) cannot be written as a critical locus.

Suppose that $M_X^{H\text{-st}}(v)$ is a fine moduli space, and $\mathcal{U} \in \text{Coh}(X \times M_X^{H\text{-st}}(v))$ is a universal sheaf. We have the natural morphism

$$\mathcal{O}_{X \times M_X^{H\text{-st}}(v)} \longrightarrow \mathbf{R}\,\mathcal{H}om_{X \times M_X^{H\text{-st}}(v)}(\mathcal{U}, \mathcal{U}).$$

We denote by $\mathbf{R}\,\mathcal{H}om_{X \times M_X^{H\text{-st}}(v)}(\mathcal{U}, \mathcal{U})_0$ the cone of the above morphism. The following result is proved by Thomas [163] and Huybrechts–Thomas [83].

Theorem 1.2 ([83, 163]) *Suppose that $M_X^{H\text{-st}}(v)$ is a fine moduli space. Then there exists a canonical symmetric perfect obstruction theory*

$$\mathcal{E} = \left(\mathbf{R}p_{M*}\mathbf{R}\,\mathcal{H}om_{X \times M_X^{H\text{-st}}(v)}(\mathcal{U}, \mathcal{U})_0[1]\right)^{\vee} \longrightarrow \tau_{\geq -1}\mathbb{L}_{M_X^{H\text{-st}}(v)}. \qquad (1.18)$$

Here $p_M \colon X \times M_X^{H\text{-st}}(v) \to M_X^{H\text{-st}}(v)$ is the projection.

Remark 1.13 It is not essential to assume that $M_X^{H\text{-st}}(v)$ is fine in Theorem 1.2. If $M_X^{H\text{-st}}(v)$ is not fine, although there is no universal sheaf, there is a twisted universal sheaf (twisted by a Brauer class, see [40]) and we can use it to construct the symmetric perfect obstruction theory (1.18). See [83, Footnote 2].

By Theorems 1.1 and 1.2, there exists a zero-dimensional virtual class

$$[M_X^{H\text{-st}}(v)]^{\text{vir}} \in A_0(M_X^{H\text{-st}}(v)).$$

Suppose furthermore that $M_X^{H\text{-st}}(v)$ satisfies the ss=st condition (see Definition 1.4). Then $M_X^{H\text{-st}}(v)$ is projective, so we can integrate the above zero-dimensional virtual class to obtain some integer. This is called the *Donaldson–Thomas (DT) invariant*.

Definition 1.7 ([163]) Suppose that X is a smooth projective CY 3-fold and $M_X^{H\text{-st}}(v)$ is fine and satisfies the ss=st condition. The DT invariant $\text{DT}_H(v) \in \mathbb{Z}$ is defined by

$$\text{DT}_H(v) := \int_{[M_X^{H\text{-st}}(v)]^{\text{vir}}} 1 \in \mathbb{Z}.$$

Example

In the setting of Definition 1.7, suppose furthermore that $M_X^{H\text{-st}}(v)$ is zero-dimensional. Then $M_X^{H\text{-st}}(v)$ is written as $\operatorname{Spec} R$ for an Artinian $\mathbb{C}$-algebra R. In this case, the DT invariant $\operatorname{DT}_H(v)$ coincides with the length of R.

Example

In the setting of Definition 1.7, suppose furthermore that $M = M_X^{H\text{-st}}(v)$ is smooth and connected. In this case, the symmetric perfect obstruction theory in Theorem 1.2 is given by

$$(T_M \xrightarrow{0} \Omega_M) \longrightarrow (0 \longrightarrow \Omega_M).$$

The virtual fundamental class associated with the above perfect obstruction theory is the Euler class of the obstruction bundle Ω_M. Therefore

$$\operatorname{DT}_H(v) = \int_{M_H(v)} e(\Omega_M) = (-1)^{\dim M} e(M).$$

Here $e(-)$ is the topological Euler number.

An important property of DT invariants is the following deformation invariance, which is proven by a general argument of perfect obstruction theory:

Theorem 1.3 ([14, Proposition 7.2]) *Let* $\pi : \mathcal{X} \to C$ *be a one-parameter family of Calabi–Yau 3-folds, and* H *a* π-*ample divisor on* $\mathcal{X}$, *and take* $v \in \Gamma(C, R\pi_*\mathbb{Q})$. *For each* $t \in C$, *suppose that* $M_{\mathcal{X}_t}^{H_t\text{-st}}(v_t)$ *is fine and satisfies the ss=st condition. Then* $\operatorname{DT}_{H_t}(v_t)$ *is independent of* $t \in C$.

Remark 1.14 It is essential to assume that X is CY3 to define DT invariants, since otherwise there may be higher obstruction spaces $\operatorname{Ext}^{\geq 3}(E, E)$ so that the obstruction theory (1.18) may not be perfect. Indeed, $\operatorname{ext}^1(E, E) - \operatorname{ext}^2(E, E)$ may not be a locally constant function of $[E] \in M_X^{H\text{-st}}(v)$ if X is not CY3. On the other hand, if X is a CY 4-fold, the Riemann–Roch theorem implies that $\operatorname{ext}^1(E, E) - \operatorname{ext}^2(E, E)/2$ is a locally constant function. Based on this observation, the virtual fundamental class of $M_X^{H\text{-st}}(v)$ for a CY 4-fold X is constructed in [23, 42, 139] by taking the half obstruction space of $\operatorname{Ext}^2(E, E)$, so that $\operatorname{ext}^1(E, E) - \operatorname{ext}^2(E, E)/2$ is the virtual dimension. The DT invariants on CY 4-folds are defined by the above virtual classes together with suitable insertions. The study of DT invariants on CY 4-folds is now developing (see [43–48]).

1.4   DT Invariants via Behrend Functions

In general, the DT invariants are not so easy to calculate from Definition 1.7. However, as we observed in the last example, the DT invariants are topological Euler numbers of the moduli spaces up to sign if they are smooth. In general, Behrend [11] proved that the DT invariants are calculated from weighted Euler characteristics on the moduli spaces of stable sheaves. This is quite useful and has led to significant developments in DT theory.

We first explain the Milnor fiber of a function on a complex manifold. Let V be a complex manifold and $f \colon V \to \mathbb{C}$ a holomorphic function. For $p \in V$, we fix a norm $\|-\|$ in a sufficiently small open neighborhood of p in V. The *Milnor fiber* of f at p is defined by

$$M_p(f) := \{x \in V : \|x - p\| < \delta, \ f(x) = f(p) + \varepsilon\} \tag{1.19}$$

for $0 < \varepsilon \ll \delta \ll 1$. The topological type of $M_p(f)$ is independent of ε, δ for $0 < \varepsilon \ll \delta \ll 1$.

For a holomorphic function $f \colon V \to \mathbb{C}$, its *critical locus* $\mathrm{Crit}(f) \subset V$ of f is defined to be the closed analytic subset defined by the ideal of $\mathcal{O}_V$ generated by the image of $df \colon T_V \to \mathcal{O}_V$. The following is the main result of [11].

Theorem 1.4 ([11, Theorem 4.18]) *For any $\mathbb{C}$-scheme M of finite type, there exists a canonical constructible function $\chi_B \colon M \to \mathbb{Z}$ satisfying the following:*

  (i) *For $x \in M$, suppose that there exists an analytic open neighborhood $x \in U \subset M$, a complex manifold $p \in V$ with a holomorphic function $f \colon V \to \mathbb{C}$ and an isomorphism $(U, x) \cong (\mathrm{Crit}(f), p)$. Then we have*

$$\chi_B(x) = (-1)^{\dim V}(1 - e(M_p(f))). \tag{1.20}$$

  (ii) *If M is proper and $\mathscr{E} \to \tau_{\geq -1}\mathbb{L}_M$ is a symmetric perfect obstruction theory, we have the identity*

$$\int_{[M]^{\mathrm{vir}}} 1 = \int_M \chi_B \, de := \sum_{k \in \mathbb{Z}} k \cdot e(\chi_B^{-1}(k)). \tag{1.21}$$

Remark 1.15 Another property of the Behrend function is that if $f \colon M_1 \to M_2$ is a smooth morphism of schemes of relative dimension d, then $f^* \chi_{B,2} = (-1)^d \chi_{B,1}$. Here $\chi_{B,i}$ is the Behrend function on M_i. Using this property, one can also define the Behrend function χ_B on any Artin stack $\mathscr{M}$, uniquely determined by the property that for any scheme M and a smooth morphism $f \colon M \to \mathscr{M}$ of relative dimension d, $(-1)^d f^* \chi_B$ is the Behrend function on M (see [92, Proposition 4.4]).

If $M = M_X^{H\text{-st}}(v)$ is a moduli space of H-stable sheaves on a CY 3-fold X satisfying the assumption in Definition 1.7, we can apply the formula (1.21) to compute the DT invariant $\mathrm{DT}_H(v)$ from the Behrend function χ_B on M. The function χ_B may be computed by Theorem 1.4 (i) and the following theorem:

Theorem 1.5 ([92, Theorem 5.4]) *If X is a CY 3-fold, the Kuranishi map in (1.11)*

$$\kappa : U \longrightarrow \mathrm{Ext}^2(E, E) \cong \mathrm{Ext}^1(E, E)^\vee$$

is written as $\kappa = df$ for some holomorphic function $f : U \to \mathbb{C}$. Here we have identified the cotangent space $\Omega_U \to U$ on U with $\mathrm{Ext}^1(E, E)^\vee \times U \to U$. In particular, the moduli space $M_X^{H\text{-st}}(v)$ is analytic locally isomorphic to $\mathrm{Crit}(f)$ on U.

The following example is due to Thomas [164].

Example

Let $X \subset \mathbb{P}^2 \times \mathbb{P}^2$ be a smooth $(3, 3)$-divisor, which is a CY 3-fold. There exists a hypersurface as above such that the first projection $X \subset \mathbb{P}^2 \times \mathbb{P}^2 \to \mathbb{P}^2$ has a singular fiber of the form $\{x^2 y = 0\} \subset \mathbb{P}^2$. Then $C = \{x = 0\} \subset X$ is isomorphic to $\mathbb{P}^1$. For $v = \mathrm{ch}(\mathcal{O}_C)$ and arbitrary polarization H, it is proved in [164, Addendum] that

$$M_X^{H\text{-st}}(v) = \mathrm{Spec}\mathbb{C}[z, w]/(z^2, w^2) = \mathrm{Crit}(f),$$

where $f : \mathbb{C}^2 \to \mathbb{C}$ is given by $f(z, w) = z^3 + w^3$. It is to easy to calculate that $e(M_0(f)) = -3$. Then we have $\mathrm{DT}_H(v) = (-1)^2(1 - (-3)) = 4$ from (1.20), (1.21) in this case. Since $M_X^{H\text{-st}}(v)$ is zero-dimensional, it is also calculated as the length of $\mathbb{C}[z, w]/(z^2, w^2)$.

1.5 Rank One DT Invariants

For $\beta \in H_2(X, \mathbb{Z})$ and $n \in \mathbb{Z}$, let $v \in \Gamma$ be an element of the form

$$v = (1, 0, -\beta, -n) \in H^0(X) \oplus H^2(X) \oplus H^4(X) \oplus H^6(X). \tag{1.22}$$

Here we have regarded β as an element of $H^4(X)$ by Poincaré duality. By the isomorphism (1.10), the moduli space $M_X^{H\text{-st}}(v)$ is fine, satisfies the ss=st condition, and it is isomorphic to the Hilbert scheme of closed subscheme $Z \subset X$ with $\dim Z \leq 1$, $[Z] = \beta$ and $\chi(\mathcal{O}_Z) = n$. The rank one DT invariant $I_{n,\beta}$ is defined by

$$I_{n,\beta} := \mathrm{DT}_H(1, 0, -\beta, -n) \in \mathbb{Z}.$$

Then we form the generating series

$$I_\beta(X) := \sum_{n \in \mathbb{Z}} I_{n,\beta} q^n, \ I(X) := \sum_{\beta \geq 0} I_\beta(X) t^\beta. \tag{1.23}$$

Here $\beta > 0$ means that β is a homology class of an effective one cycle on X. It is easy to see that $I_{n,\beta} = 0$ for $n \ll 0$ (see [168, Lemma 3.10]), so that $I_\beta(X)$ is a formal Laurent series.

Example

Suppose that $\beta = 0$. Then $M_X^{H\text{-st}}(v)$ is isomorphic to the Hilbert scheme of n-points in X. In this case, the series $I_{\beta=0}(X)$ is computed in [15, 113, 117] (conjectured in [126, Conjecture 1]):

$$I_{\beta=0}(X) = M(-q)^{e(X)}. \tag{1.24}$$

Here $M(q)$ is defined by

$$M(q) = \prod_{k \geq 1} \frac{1}{(1-q^k)^k} = 1 + q + 3q^2 + 6q^3 + \cdots.$$

Remark 1.16 The coefficient of q^d of the function $M(q)$ is the number of plane partitions of size d, and is known as the *MacMahon function*. Here we see a relationship between DT theory and enumerative combinatorics [156].

The following example gives a computation of DT invariants for the simplest curve inside CY 3-folds.

Example

Let $C \subset X$ be a curve such that $C \cong \mathbb{P}^1$ with a normal bundle isomorphic to $\mathcal{O}_{\mathbb{P}^1}(-1)^{\oplus 2}$. Suppose that any curve in X with a homology class proportional to C is supported on C. Then we have the following (see [12, Theorem 2.14]):

$$\sum_{d \geq 0} I_{d[C]}(X) t^d = M(-q)^{e(X)} \cdot \prod_{k \geq 1} (1 - (-q)^k t)^k. \tag{1.25}$$

In particular, for the single curve class $[C]$, combined with (1.24) we have

$$\frac{I_{[C]}(X)}{I_0(X)} = q - 2q^2 + 3q^3 - \cdots = \frac{q}{(1+q)^2}. \tag{1.26}$$

Note that the right hand side of (1.26) is a rational function of q, invariant under $q \mapsto q^{-1}$. Indeed, this property was conjectured in [126, Conjecture 2] to be true in general, and proved using wall-crossing in the derived category.

Theorem 1.6 ([31, 169, 170]) *The quotient series*

$$\frac{I_\beta(X)}{I_0(X)}$$

is the Laurent expansion of a rational function of q, invariant under $q \mapsto q^{-1}$.

The generating series (1.23) first appeared in the MNOP paper [126] in order to formulate the famous GW/DT correspondence conjecture. Here we briefly review the GW theory (see [10, 50, 118] for details). By definition, the data $(C, p_1, \ldots, p_n, f)$ is called a *stable map* to a smooth projective variety X if:

(i) C is a projective curve with at worst nodal singularities and $p_1, \ldots, p_n \in C$ lie on the smooth locus of C.
(ii) $f : C \to X$ is a morphism such that the group of automorphisms of C preserving f and p_i is a finite group.

For each $g \geq 0$ and $\beta \in H_2(X, \mathbb{Z})$, the moduli space $\overline{M}_{g,n}(X, \beta)$ of stable maps $(C, p_1, \ldots, p_n, f)$ with $g(C) = g$, $f_*[C] = \beta$ is a Deligne–Mumford stack with a natural perfect obstruction theory. Its virtual dimension is

$$\dim[\overline{M}_{g,n}(X, \beta)]^{\text{vir}} = -K_X \cdot \beta + (\dim X - 3)(1 - g) + n.$$

In particular, if $n = 0$ and X is a CY 3-fold, then $\overline{M}_{g,0}(X, \beta)$ has virtual dimension zero. The corresponding *Gromov–Witten (GW) invariant* is defined by

$$\mathrm{GW}_{g,\beta} := \int_{[\overline{M}_{g,0}(X,\beta)]^{\text{vir}}} 1 \in \mathbb{Q}.$$

Conjecture 1.1 ([126, Conjecture 3]) After the variable change $q = -e^{i\lambda}$, we have the identity

$$\exp\left(\sum_{g \geq 0, \beta > 0} \mathrm{GW}_{g,\beta} \lambda^{2g-2} t^\beta \right) = \frac{I(X)}{I_0(X)}. \tag{1.27}$$

Here we need some more explanation of the identity (1.27). By Theorem 1.6, the right hand side of (1.27) is a rational function of q, so it can be analytically continued from $q = 0$ to $q = -1$. The identity (1.27) should hold after taking the above analytic continuation and expanding it around $q = -1$ via the variable change $q = -e^{i\lambda}$ (so expanding it at $\lambda = 0$). For example, in the case of (1.26), we have

$$\frac{q}{(1+q)^2}\bigg|_{q=-e^{i\lambda}} = -\frac{e^{i\lambda}}{(1-e^{i\lambda})^2} = \left(2\sin\left(\frac{\lambda}{2}\right)\right)^{-2}.$$

The identity in Conjecture 1.1 is proved in many classes of CY 3-folds including quintic 3-folds by Pandharipande–Pixton [145]. At this moment, the following techniques are the fundamental tools to approach Conjecture 1.1:

- Use Li's degeneration formula [116, 119] for GW/DT theory to reduce the desired identity to a toric case.
- Use the Graber–Pandharipande virtual localization formula [73] for the torus action on toric 3-folds.

Although both sides of (1.27) count curves inside a CY 3-fold X, the geometric meanings of both countings are quite different. On the GW side the source curve C has at most nodal singularities, while on the DT side there are no constraints of curves. On the GW side the map $f : C \to X$ is not necessarily an embedding, while on the DT side the curve is embedded into X. Also the GW invariants are not necessarily integers due to the multiple cover phenomena of the map $f : C \to X$, while the DT invariants are always integers as the curves are always embedded. In particular, the correspondence (1.27) implies some hidden integrality property of GW invariants.

In Chap. 7, we will discuss another hidden integrality property of GW invariants in terms of Gopakumar–Vafa invariants.

Chapter 2
Generalized Donaldson–Thomas Invariants

Abstract In the last chapter, we gave a definition of DT invariants on CY 3-folds when the moduli spaces of stable sheaves are fine and satisfy the ss=st condition. Although the first condition is not essential, the latter condition is much more essential, and it is much more difficult to define DT invariants when there exist strictly semistable sheaves. In this chapter, we explain the construction of DT invariants without the ss=st condition by Joyce–Song, using motivic Hall algebras.

2.1 An Idea for Generalized DT Theory

In Chap. 1, we defined the DT invariants on CY 3-folds via integrals of the virtual classes under the assumption that the moduli space of stable sheaves $M_X^{H\text{-st}}(v)$ is fine and satisfies the ss=st condition (see Definition 1.7). Although the first condition is not essential (see Remark 1.13), the latter condition is essential by the following reasons:

- If $M_X^{H\text{-st}}(v)$ does not satisfy the ss=st condition, then it may not be proper. So we cannot integrate the virtual class as in Definition 1.7.
- In order to construct a virtual class on a proper moduli space, one may try to extend the obstruction theory (1.18) to the strictly semistable locus. However, for a strictly semistable sheaf E, the complex $\mathbf{R}\mathrm{Hom}(E, E)_0$ is four term so we cannot construct a perfect obstruction theory as in Theorem 1.2.

Alternatively, using Theorem 1.4 one may try to define DT invariants via Behrend functions when $M_X^{H\text{-st}}(v)$ does not satisfy the ss=st condition. However, we also have the following issues:

- If we define the DT invariants as integrals of Behrend functions on $M_X^{H\text{-st}}(v)$, then they are not necessarily deformation invariant (see [92, Corollary 6.22]).
- One may try a similar approach for the good moduli space $M_X^{H\text{-ss}}(v)$, which is proper. However, it is not locally written as a critical locus, so there is no reason to consider the Behrend function on it.

© The Author(s), under exclusive license to Springer Nature Singapore Pte Ltd. 2021   17
Y. Toda, *Recent Progress on the Donaldson—Thomas Theory*,
SpringerBriefs in Mathematical Physics 43,
https://doi.org/10.1007/978-981-16-7838-7_2

An approach taken by Joyce–Song [92] is to consider Behrend functions on the stack $\mathscr{M}_X^{H\text{-ss}}(v)$. The Behrend functions on Artin stacks make sense by Remark 1.15, and that on $\mathscr{M}_X^{H\text{-ss}}(v)$ is reasonable to consider since it is locally a critical locus (see [92, Theorem 5.5], [18, Theorem 1.2]). However, it is not obvious how to take its weighted Euler characteristics. For example, let us take $v = (2, 0, 0, 0)$ so that $\mathscr{M}_X^{H\text{-ss}}(v)$ consists of $\{\mathcal{O}_X^{\oplus 2}\}$. Then the moduli stack $\mathscr{M}_X^{H\text{-ss}}(v)$ is $BGL_2(\mathbb{C})$. We may try to think of the corresponding DT invariant as something like the Euler number of $\mathscr{M}_X^{H\text{-ss}}(v) = BGL_2(\mathbb{C})$, or that of its $\mathbb{C}^*$-rigidification $BPGL_2(\mathbb{C})$. In both cases, a naive guess for their Euler numbers

$$\frac{1}{e(GL_2(\mathbb{C}))}, \quad \frac{1}{e(PGL_2(\mathbb{C}))}$$

does not make sense as the denominators are zero.

An issue caused here is that, for a strictly semistable sheaf E, the dimension of the maximal torus of $\mathrm{Aut}(E)$ is greater than one (in the above case, the maximal torus of $GL_2(\mathbb{C})$ is $(\mathbb{C}^*)^2$). A crucial idea in Joyce's series of papers [86–89] is that, by taking the 'logarithm' of the moduli stack $\mathscr{M}_X^{H\text{-ss}}(v)$, we can regard it as a 'virtual' stack whose stabilizer groups have one-dimensional maximal torus. Then by getting rid of the one-dimensional maximal torus and integrating over the Behrend function, the desired DT invariant (called *generalized DT invariant*) is defined.

The process of taking the 'logarithm' can be made precise using motivic Hall algebras of coherent sheaves. The motivic Hall algebra is also important in the wall-crossing formula of DT invariants.

2.2 Motivic Hall Algebras

We define motivic Hall algebras following the convention of Bridgeland [26, Sect. 4], which simplifies Joyce's version [87, Sect. 5]. Let $\mathscr{S}$ be an Artin stack locally of finite type with affine geometric stabilizers. We first define the *Grothendieck group of stacks* over $\mathscr{S}$.

Definition 2.1 ([26, Definition 3,10]) We define $K(\mathrm{St}/\mathscr{S})$ to be the $\mathbb{Q}$-vector space spanned by isomorphism classes of symbols $[\mathscr{X} \xrightarrow{\rho} \mathscr{S}]$, where $\mathscr{X}$ is an Artin stack with affine geometric stabilizers. The relations are generated by the following:

(i) For every pair of $\mathscr{X}_1, \mathscr{X}_2$, we have

$$[\mathscr{X}_1 \sqcup \mathscr{X}_2 \xrightarrow{\rho_1 \sqcup \rho_2} \mathscr{S}] = [\mathscr{X}_1 \xrightarrow{\rho_1} \mathscr{S}] + [\mathscr{X}_2 \xrightarrow{\rho_2} \mathscr{S}].$$

(ii) For every morphism $\rho \colon \mathscr{X}_1 \to \mathscr{X}_2$ such that $\mathscr{X}_1(\mathbb{C}) \to \mathscr{X}_2(\mathbb{C})$ is an equivalence, and a morphism $\rho' \colon \mathscr{X}_2 \to \mathscr{S}$, we have $[\mathscr{X}_1 \xrightarrow{\rho \circ \rho'} \mathscr{S}] = [\mathscr{X}_2 \xrightarrow{\rho'} \mathscr{S}]$.

(iii) Let $h_i\colon \mathcal{X}_i \to \mathcal{Y}$ for $i = 1, 2$ satisfy the following: for any scheme T and a morphism $T \to \mathcal{Y}$, the pull-backs $\mathcal{X}_i \times_{\mathcal{Y}} T \to T$ are Zariski locally trivial with the same fiber F. Then we have $[\mathcal{X}_1 \overset{goh_1}{\to} \mathcal{S}] = [\mathcal{X}_2 \overset{goh_2}{\to} \mathcal{S}]$.

When $\mathcal{S} = \mathrm{Spec}\,\mathbb{C}$, we write $K(\mathrm{St}) := K(\mathrm{St}/\mathrm{Spec}\,\mathbb{C})$. There is an algebra structure on $K(\mathrm{St})$ given by the product of stacks, $[\mathcal{X}_1] \cdot [\mathcal{X}_2] = [\mathcal{X}_1 \times \mathcal{X}_2]$. Let $K(\mathrm{Var}) \subset K(\mathrm{St})$ be the subalgebra spanned by the classes of varieties. We also denote by $\mathbb{L} \in K(\mathrm{Var})$ the class of the affine line $\mathbb{A}^1$. It is proved in [26, Lemma 3.9] that we have the identity

$$K(\mathrm{St}) = K(\mathrm{Var})\left[\frac{1}{\mathbb{L}}, \frac{1}{\mathbb{L}-1}, \frac{1}{\mathbb{L}^n + \cdots + \mathbb{L}+1} : n \geq 1\right]. \qquad (2.1)$$

Let X be a smooth projective CY 3-fold, and $\mathcal{M}_X$ the stack of coherent sheaves on X, i.e. disjoint union of stacks $\mathcal{M}_X(v)$ in (1.3) for all $v \in \Gamma$. The motivic Hall-algebra of coherent sheaves on X is defined as follows:

Definition 2.2 We define the *motivic Hall algebra* $H(\mathrm{Coh}(X))$ by

$$H(\mathrm{Coh}(X)) := K(\mathrm{St}/\mathcal{M}_X).$$

Note that $H(\mathrm{Coh}(X))$ is naturally a module over $K(\mathrm{St})$ by

$$[\mathcal{X}] \cdot [\mathcal{Y} \overset{\rho}{\to} \mathcal{M}_X] = [\mathcal{X} \times \mathcal{Y} \overset{p}{\to} \mathcal{Y} \overset{\rho}{\to} \mathcal{M}_X],$$

where p is the projection. There is an associative $K(\mathrm{St})$-algebra structure on $H(\mathrm{Coh}(X))$ based on the *Ringel–Hall algebra* [153]. Let $\mathcal{M}_X^{\mathrm{ex}}$ be the 2-functor

$$\mathcal{M}_X^{\mathrm{ex}}\colon (Sch/\mathbb{C}) \longrightarrow (Groupoid)$$

sending a $\mathbb{C}$-scheme T to the groupoid of exact sequences in $\mathrm{Coh}(X \times T)$

$$0 \longrightarrow \mathcal{E}_1 \longrightarrow \mathcal{E}_2 \longrightarrow \mathcal{E}_3 \longrightarrow 0, \qquad (2.2)$$

where each $\mathcal{E}_i \in \mathrm{Coh}(X \times T)$ is flat over T. The isomorphisms of the above groupoid consist of commutative isomorphisms

$$
\begin{array}{ccccccccc}
0 & \longrightarrow & \mathcal{E}_1 & \longrightarrow & \mathcal{E}_2 & \longrightarrow & \mathcal{E}_3 & \longrightarrow & 0 \\
& & \downarrow \cong & & \downarrow \cong & & \downarrow \cong & & \\
0 & \longrightarrow & \mathcal{E}_1' & \longrightarrow & \mathcal{E}_2' & \longrightarrow & \mathcal{E}_3' & \longrightarrow & 0.
\end{array}
$$

The stack $\mathcal{M}_X^{\mathrm{ex}}$ is an Artin stack locally of finite type (see [26, Lemma 4.1]), which admits morphisms

$$p_i : \mathcal{M}_X^{\mathrm{ex}} \longrightarrow \mathcal{M}_X, \ \mathcal{E}_\bullet \longmapsto \mathcal{E}_i.$$

Then there is an $*$-product on $H(\mathrm{Coh}(X))$ given by

$$[\mathcal{X}_1 \xrightarrow{\rho_1} \mathcal{M}_X] * [\mathcal{X}_2 \xrightarrow{\rho_2} \mathcal{M}_X] = [\mathcal{X}_3 \xrightarrow{\rho_3} \mathcal{M}_X]$$

where $(\mathcal{X}_3, \rho_3 = p_3 \circ \eta)$ is given by the following diagram

$$
\begin{array}{ccccc}
\mathcal{X}_3 & \xrightarrow{\ \eta\ } & \mathcal{M}_X^{\mathrm{ex}} & \xrightarrow{\ p_3\ } & \mathcal{M}_X \\
\downarrow & & \square & \downarrow {\scriptstyle (p_1,p_2)} & \\
\mathcal{X}_1 \times \mathcal{X}_2 & \xrightarrow{(\rho_1,\rho_2)} & \mathcal{M}_X^{\times 2}. & &
\end{array}
$$

The unit is given by $1 = [\mathrm{Spec}\,\mathbb{C} \to \mathcal{M}_X]$ which corresponds to $0 \in \mathrm{Coh}(X)$.

Recall that Γ was defined by the image of the Chern character map (1.2). The algebra $H(\mathrm{Coh}(X))$ is Γ-graded,

$$H(\mathrm{Coh}(X)) = \bigoplus_{v \in \Gamma} H_v(\mathrm{Coh}(X)),$$

where $H_v(\mathrm{Coh}(X))$ consists of elements $[\mathcal{X} \xrightarrow{\rho} \mathcal{M}_X]$ such that ρ factors through $\mathcal{M}_X(v) \subset \mathcal{M}_X$.

Example

For $E \in \mathrm{Coh}(X)$, the *'delta function'* at E is defined by

$$\delta_E := [\mathrm{Spec}\,\mathbb{C} \longrightarrow \mathcal{M}_X] \in H(\mathrm{Coh}(X)), \ \bullet \longmapsto [E].$$

Then the $*$-product of the delta functions δ_{E_i} for $i = 1, 2$ is calculated as follows (see [30, Proposition 6.2]):

$$\delta_{E_1} * \delta_{E_2} = \left[\left[\mathrm{Ext}^1(E_2, E_1)/\mathrm{Hom}(E_2, E_1) \right] \xrightarrow{\rho} \mathcal{M}_X \right] \tag{2.3}$$

$$= \mathbb{L}^{-\hom(E_2, E_1)} \left[\mathrm{Ext}^1(E_2, E_1) \xrightarrow{\rho} \mathcal{M}_X \right]. \tag{2.4}$$

Here $\mathrm{Hom}(E_2, E_1)$ acts on $\mathrm{Ext}^1(E_2, E_1)$ trivially; the map ρ sends $\xi \in \mathrm{Ext}^1(E_2, E_1)$ to $[E_\xi] \in \mathcal{M}_X$ corresponding to the extension class $0 \to E_1 \to E_\xi \to E_2 \to 0$ of ξ.

2.3 Poisson Algebra Homomorphism

In this section, we construct two Poisson algebras and give a statement relating them. Here is a general definition of Poisson algebras.

Definition 2.3 Let $(V, *, \{-.-\})$ be a triple such that V is a $\mathbb{Q}$-vector space and $*$, $\{-, -\}$ are bilinear forms $V \times V \to \mathbb{Q}$. The above triple is called a *Poisson algebra* if the following conditions hold:

  (i) $(V, *)$ is an associative algebra.
 (ii) $(V, \{-, -\})$ is a Lie algebra. The Lie bracket $\{-, -\}$ is called a *Poisson bracket*.
(iii) The Poisson bracket $\{-, -\}$ is a derivation with respect to $*$, i.e. for any $x, y, z \in V$ we have $\{x, y * z\} = \{x, y\} * z + y * \{x, z\}$.

Let Λ be the subalgebra of $K(\mathrm{St})$ defined by (see the identity (2.1))

$$\Lambda := K(\mathrm{Var}) \left[\frac{1}{\mathbb{L}}, \frac{1}{\mathbb{L}^n + \cdots + \mathbb{L} + 1} : n \geq 1 \right] \subset K(\mathrm{St}).$$

The Λ-submodule

$$H^{\mathrm{reg}}(\mathrm{Coh}(X)) \subset H(\mathrm{Coh}(X))$$

is defined to be spanned by $[Z \to \mathscr{M}_X]$ so that Z is a variety.

Remark 2.1 A point of defining $H^{\mathrm{reg}}(\mathrm{Coh}(X))$ is that one can define the Euler number for any element in $H^{\mathrm{reg}}(\mathrm{Coh}(X))$. Indeed, there is a unique map

$$e \colon H^{\mathrm{reg}}(\mathrm{Coh}(X)) \longrightarrow \mathbb{Q} \tag{2.5}$$

sending $[Z \to \mathscr{M}_X]$ for a variety Z to $e(Z)$. This is possible since $e(\mathbb{A}^1) = 1$ and the denominators of the fractions in Λ do not have poles at $\mathbb{L} = 1$.

The $*$-product on $H(\mathrm{Coh}(X))$ preserves $H^{\mathrm{reg}}(\mathrm{Coh}(X))$, i.e. $H^{\mathrm{reg}}(\mathrm{Coh}(X))$ is a subalgebra of $H(\mathrm{Coh}(X))$. Indeed, the product of the delta-functions (2.3) is certainly an element in $H^{\mathrm{reg}}(\mathrm{Coh}(X))$. Moreover, the quotient algebra

$$H^{\mathrm{sc}}(\mathrm{Coh}(X)) := H^{\mathrm{reg}}(\mathrm{Coh}(X))/(\mathbb{L} - 1)H^{\mathrm{reg}}(\mathrm{Coh}(X))$$

is a commutative algebra (see [26, Theorem 5.1]). Therefore for $f, g \in H^{\mathrm{sc}}(\mathrm{Coh}(X))$, we can define the following bracket on $H^{\mathrm{sc}}(\mathrm{Coh}(X))$:

$$\{f, g\} := \frac{f * g - g * f}{\mathbb{L} - 1}. \tag{2.6}$$

We note that (2.6) is well-defined since $\mathbb{L} - 1 = [\mathbb{C}^*]$ is invertible in $K(\mathrm{St})$. By the $*$-product together with the above bracket $\{-, -\}$, we have the Poisson algebra structure on $H^{\mathrm{sc}}(\mathrm{Coh}(X))$.

Example

In the case of delta-functions δ_{E_i}, by (2.3) we have the identity in $H^{sc}(\text{Coh}(X))$

$$\delta_{E_1} * \delta_{E_2} = \mathbb{L}^{-\hom(E_2,E_1)}\delta_{E_1 \oplus E_2} + \mathbb{L}^{-\hom(E_2,E_1)}(\mathbb{L}-1)[\mathbb{P}(\text{Ext}^1(E_2,E_1)) \longrightarrow \mathcal{M}_X]$$
$$= \mathbb{L}^{-\hom(E_2,E_1)}\delta_{E_1 \oplus E_2}. \tag{2.7}$$

Then we have

$$\delta_{E_1} * \delta_{E_2} - \delta_{E_2} * \delta_{E_1} = (\mathbb{L}^{-\hom(E_2,E_1)} - \mathbb{L}^{-\hom(E_1,E_2)}) \cdot \delta_{E_1 \oplus E_2} = 0$$

since $(\mathbb{L}^{-\hom(E_2,E_1)} - \mathbb{L}^{-\hom(E_1,E_2)})$ is divisible by $(\mathbb{L}-1)$. The Poisson bracket (2.6) is given by

$$\{\delta_{E_1}, \delta_{E_2}\} = \frac{(\mathbb{L}^{-\hom(E_2,E_1)} - \mathbb{L}^{-\hom(E_1,E_2)})}{\mathbb{L}-1} \cdot \delta_{E_1 \oplus E_2}$$
$$+ \mathbb{L}^{-\hom(E_2,E_1)}[\mathbb{P}(\text{Ext}^1(E_2,E_1)) \longrightarrow \mathcal{M}_X]$$
$$- \mathbb{L}^{-\hom(E_1,E_2)}[\mathbb{P}(\text{Ext}^1(E_1,E_2)) \longrightarrow \mathcal{M}_X]. \tag{2.8}$$

Let us take $v_i \in \Gamma$ for $i = 1, 2$ and write them as

$$v_i = (r_i, D_i, -\beta_i, -n_i) \in H^0(X) \oplus H^2(X) \oplus H^4(X) \oplus H^6(X).$$

We define the bilinear form $\chi : \Gamma \times \Gamma \to \mathbb{Z}$ by

$$\chi(v_1, v_2) = n_1 r_2 - n_2 r_1 + D_1 \beta_2 - D_2 \beta_1 + \frac{c_2(X)}{12}(r_1 D_2 - r_2 D_1). \tag{2.9}$$

This bilinear form is determined so that we have the following identities for $E_1, E_2 \in \text{Coh}(X)$,

$$\chi(E_1, E_2) := \sum_{i \in \mathbb{Z}}(-1)^i \text{ext}^i(E_1, E_2)$$
$$= \hom(E_1, E_2) - \text{ext}^1(E_1, E_2) + \text{ext}^1(E_2, E_1) - \hom(E_2, E_1)$$
$$= \chi(v(E_1), v(E_2)).$$

Here the second identity follows from Serre duality and the last identity follows from the Riemann–Roch theorem.

We define another Poisson algebra $C(X)$ (called *Poisson torus*) by

$$C(X) := \bigoplus_{v \in \Gamma} \mathbb{Q} \cdot c_v.$$

Namely as a $\mathbb{Q}$-vector space, it is spanned by all the elements in Γ. The $*$-product on $C(X)$ is defined by

$$c_{v_1} * c_{v_2} := (-1)^{\chi(v_1, v_2)} c_{v_1 + v_2}.$$

The Poisson bracket is defined by

$$\{c_{v_1}, c_{v_2}\} := (-1)^{\chi(v_1, v_2)} \chi(v_1, v_2) c_{v_1 + v_2}.$$

It is easy to see that the above defined product and the Poisson bracket give the Poisson algebra structure on $C(X)$.

The following theorem (which simplifies [87, Theorem 6.12]) is a fundamental result on the theory of motivic Hall algebras which relates the Poisson algebra $H^{sc}(\mathrm{Coh}(X))$ with the Poisson torus $C(X)$.

Theorem 2.1 ([26, Theorem 5.2]) *There is a Γ-graded Poisson algebra homomorphism*

$$I: H^{sc}(\mathrm{Coh}(X)) \longrightarrow C(X)$$

such that for a variety Z with a morphism $\rho: Z \to \mathcal{M}_X(v)$, we have

$$I([Z \xrightarrow{\rho} \mathcal{M}_X(v)]) = \left(\int_Z \rho^* \chi_B \, de \right) \cdot c_v. \qquad (2.10)$$

Here χ_B is the Behrend function on the stack $\mathcal{M}_X(v)$ (see Remark 1.15).

Remark 2.2 A simpler version of Theorem 2.1 also holds by formally setting $\chi_B \equiv 1$, and modifying the $*$-product and the Poisson bracket on $C(X)$ by

$$c_{v_1} *' c_{v_2} = c_{v_1 + v_2}, \quad \{c_{v_1}, c_{v_2}\}' = \chi(v_1, v_2) c_{v_1 + v_2}.$$

Example

Let us check a simpler version in Remark 2.2 for delta-functions δ_{E_i} for $E_i \in \mathrm{Coh}(X)$ with $v_i = \mathrm{ch}(E_i)$. We denote by I' the one defined as (2.10) by formally setting $\chi_B \equiv 1$. From (2.7), we have

$$I'(\delta_{E_1} * \delta_{E_2}) = I'(\mathbb{L}^{-\hom(E_2, E_1)} \delta_{E_1 \oplus E_2})$$
$$= c_{v_1 + v_2}$$
$$= c_{v_1} *' c_{v_2}.$$

From (2.8), we have

$$I'(\{\delta_{E_1}, \delta_{E_2}\}) = (\hom(E_1, E_2) - \hom(E_2, E_1) - \mathrm{ext}^1(E_1, E_2) + \mathrm{ext}^1(E_2, E_1))c_{v_1+v_2}$$
$$= \chi(v_1, v_2)c_{v_1+v_2}$$
$$= \{c_{v_1}, c_{v_2}\}'. \tag{2.11}$$

2.4 Definition of Generalized DT Invariants

Let us take an ample divisor H on X and $v \in \Gamma$. Then as in (1.4), the moduli stack of H-semistable sheaves $\mathcal{M}_X^{H\text{-ss}}(v)$ is an open substack of $\mathcal{M}_X(v)$ of finite type. Therefore it determines an element

$$\delta^{H\text{-ss}}(v) := [\mathcal{M}_X^{H\text{-ss}}(v) \subset \mathcal{M}_X(v)] \in H_v(\mathrm{Coh}(X)).$$

The 'logarithm' of the above moduli stack can be defined using the motivic Hall algebra.

Definition 2.4 ([89, Definition 3.18]) We define $\epsilon^{H\text{-ss}}(v) \in H_v(\mathrm{Coh}(X))$ by

$$\epsilon^{H\text{-ss}}(v) := \sum_{k \geq 1} \sum_{\substack{v_1+\cdots+v_k=v \\ \overline{\chi}(v_i(m))=\overline{\chi}(v(m))}} \frac{(-1)^{k-1}}{k} \delta^{H\text{-ss}}(v_1) * \cdots * \delta^{H\text{-ss}}(v_k). \tag{2.12}$$

Here $\overline{\chi}(v(m))$ is the reduced Hilbert polynomial $\overline{\chi}(E(m))$ for $E \in \mathrm{Coh}(X)$ with $\mathrm{ch}(E) = v$.

Remark 2.3 It is easy to see that the right hand side of (2.12) is a finite sum, so that $\epsilon^{H\text{-ss}}(v)$ is well-defined. The reason we may regard $\epsilon^{H\text{-ss}}(v)$ as a logarithm is that, in some suitable completion of $H(\mathrm{Coh}(X))$, we have the following identity for each fixed $p(m) \in \mathbb{Q}[m]$:

$$\sum_{v \in \Gamma, \overline{\chi}(v(m))=p(m)} \epsilon^{H\text{-ss}}(v) = \log\left(1 + \sum_{v \in \Gamma, \overline{\chi}(v(m))=p(m)} \delta^{H\text{-ss}}(v)\right).$$

Here the right hand side is calculated by the Taylor expansion $\log(1 + x) = \sum_{k \geq 1} (-1)^{k-1} x^k / k$.

The following is a very deep theorem by Joyce, which roughly implies that the 'virtual' stack $\epsilon^{H\text{-ss}}(v)$ has only one-dimensional maximal tori in stabilizer groups.

The proof by Joyce is very complicated, and we also refer to another approach by Behrend–Ronagh [16] using inertia operators on stacks.

Theorem 2.2 ([88, Theorem 8.7]) *We have*

$$(\mathbb{L} - 1) \cdot \epsilon^{H\text{-ss}}(v) \in H^{\text{reg}}(\text{Coh}(X)).$$

By Theorem 2.2, one can define the element

$$\widehat{\epsilon}^{H\text{-ss}}(v) := [(\mathbb{L} - 1) \cdot \epsilon^{H\text{-ss}}(v)] \in H^{\text{sc}}(\text{Coh}(X)).$$

The generalized DT invariants are defined by applying the Poisson algebra homomorphism in Theorem 2.1.

Definition 2.5 ([92, Definition 5.15]) The *generalized DT invariant* $\text{DT}_H(v) \in \mathbb{Q}$ is defined by the identity

$$I(\widehat{\epsilon}^{H\text{-ss}}(v)) = -\text{DT}_H(v) \cdot c_v. \tag{2.13}$$

Remark 2.4 If $M_X^{H\text{-st}}(v)$ satisfies the assumption in Definition 1.7, then $\epsilon^{H\text{-ss}}(v) = \delta^{H\text{-ss}}(v)$ holds. In this case, the invariant $\text{DT}_H(v)$ coincides with the one defined in Definition 1.7 by the property of the Behrend function (1.21). The reason for the minus sign in the right hand side of (2.5) is that the Behrend function on the stack $\mathcal{M}_X^{H\text{-st}}(v)$ has the opposite sign to that on $M_X^{H\text{-st}}(v)$.

Example

Let us take $v_n = n \cdot \text{ch}(\mathcal{O}_X)$ for $n \geq 1$. Then $\mathcal{O}_X^{\oplus n}$ is a H-semistable sheaf with Chern character v_n for any polarization H. As we assume $H^1(\mathcal{O}_X) = 0$, we have $\text{Ext}^1(\mathcal{O}_X^{\oplus n}, \mathcal{O}_X^{\oplus n}) = 0$, so $\mathcal{O}_X^{\oplus n}$ is an isolated point in the moduli stack $\mathcal{M}_X^{H\text{-ss}}(v_n)$. Together with $\text{Aut}(\mathcal{O}_X^{\oplus n}) = \text{GL}_n(\mathbb{C})$, we see that $\mathcal{M}_X^{H\text{-ss}}(v_n)$ contains $B\text{GL}_n(\mathbb{C})$ as a connected component. We assume that $\mathcal{M}_X^{H\text{-ss}}(v_n) = B\text{GL}_n(\mathbb{C})$ holds and compute the generalized DT invariant in this case. In the $n = 1$ case, then $M_X^{H\text{-st}}(v)$ satisfies the assumption in Definition 1.7 and $\text{DT}_H(v_1) = 1$. Below we consider the $n = 2$ case.

The element $\epsilon^{H\text{-ss}}(v_2)$ is calculated as

$$\epsilon^{H\text{-ss}}(v_2) = [B\text{GL}_2(\mathbb{C})] \cdot \delta_{\mathcal{O}_X^{\oplus 2}} - \frac{1}{2}((B\mathbb{C}^* \cdot \delta_{\mathcal{O}_X}) * (B\mathbb{C}^* \cdot \delta_{\mathcal{O}_X}))$$

$$= \left(B\text{GL}_2(\mathbb{C}) - \frac{1}{2}\mathbb{L}^{-1}(B\mathbb{C}^*)^{\times 2}\right) \cdot \delta_{\mathcal{O}_X^{\oplus 2}}$$

$$= \left(\frac{1}{\mathbb{L}(\mathbb{L} - 1)(\mathbb{L}^2 - 1)} - \frac{1}{2\mathbb{L}(\mathbb{L} - 1)^2}\right) \cdot \delta_{\mathcal{O}_X^{\oplus 2}}$$

$$= -\frac{1}{2\mathbb{L}(\mathbb{L}+1)(\mathbb{L}-1)}\delta_{\mathcal{O}_X^{\oplus 2}}.$$

Here we have used $[\mathbb{C}^*] = \mathbb{L} - 1$, $[GL_2(\mathbb{C})] = \mathbb{L}(\mathbb{L}-1)(\mathbb{L}^2-1)$, and $\delta_{\mathcal{O}_X} * \delta_{\mathcal{O}_X} = \mathbb{L}^{-1}\delta_{\mathcal{O}_X^{\oplus 2}}$ from (2.3). It follows that we have

$$\widehat{\epsilon}^{H\text{-ss}}(v_2) = -\frac{1}{2\mathbb{L}(\mathbb{L}+1)}\delta_{\mathcal{O}_X^{\oplus 2}} = -\frac{1}{4}\delta_{\mathcal{O}_X^{\oplus 2}}.$$

Since the Behrend function on $\mathcal{O}_X^{\oplus 2}$ is 1, we conclude that $DT_H(v_2) = 1/4$. For any $n \geq 2$, it is known that $DT_H(v_n) = 1/n^2$ (see [92, Example 6.2]).

The generalized DT invariant $DT_H(v) \in \mathbb{Q}$ is not an integer in general. Instead we define $\Omega_H(v) \in \mathbb{Q}$ by the identity

$$DT_H(v) = \sum_{k \geq 1, k|v} \frac{1}{k^2}\Omega_H(v/k). \tag{2.14}$$

Note that $\Omega_H(v)$ is defined inductively by the divisibility of $v \in \Gamma$. The invariant $\Omega_H(v)$ is called *BPS invariant*, and expected to be an integer under some condition.

Conjecture 2.1 ([92, Conjecture 6.12], [108, Conjecture 6]) Suppose that for any $v' \in \Gamma$ with $\overline{\chi}(v', m) = \overline{\chi}(v, m)$, we have $\chi(v', v) = 0$. Then $\Omega_H(v) \in \mathbb{Z}$.

2.5　Joyce–Song Pair Invariants

In general, it is not easy to compute generalized DT invariants from Definition 2.5. In [92], Joyce–Song introduced another invariants (called Joyce–Song (JS) pair invariants) which can be defined from fine moduli spaces satisfying the ss=st condition, and relate them with generalized DT invariants via wall-crossing. Such a comparison is quite useful to compute generalized DT invariants in practice. Let $\mathcal{O}_X(1)$ be an ample line bundle on X with $c_1(\mathcal{O}_X(1)) = H$. The JS pair is defined as follows:

Definition 2.6 ([92, Definition 5.24]) A pair (E, s) for $E \in \text{Coh}(X)$ and $0 \neq s \in H^0(E(n))$ for $n \in \mathbb{Z}$ is called a *Joyce–Song (JS) pair* if the following conditions are satisfied:

(i) E is a H-semistable sheaf;
(ii) for any subsheaf $E' \subsetneq E$ such that $s \in H^0(E'(n))$, we have $\overline{\chi}(E', m) \prec \overline{\chi}(E, m)$.

For each $v \in \Gamma$ and $n \in \mathbb{Z}$, we denote by $M_X^{JS}(v, n)$ the moduli space of JS pairs (E, s) for $s \in H^0(E(n))$ such that $\text{ch}(E) = v$. In [92, Theorem 5.22], it is proved that $M_X^{JS}(v, n)$ is a projective scheme. Moreover, for $n \gg 0$, the JS pair moduli space $M_X^{JS}(v, n)$ admits a symmetric perfect obstruction theory (see [92, Theorem 5.23]). The JS pair invariant is defined as follows:

Definition 2.7 For $n \gg 0$, the *Joyce–Song (JS) pair invariant* is defined by

$$JS(v, n) := \int_{[M_X^{JS}(v,n)]^{\mathrm{vir}}} 1 \in \mathbb{Z}.$$

By Theorem 1.4, the invariant $JS(v, n)$ equals the weighted Euler characteristic of the Behrend function on $M_X^{JS}(v, n)$. The following theorem proved by Joyce–Song gives a comparison of JS pair invariants with generalized DT invariants.

Theorem 2.3 ([92, Proposition 5.29]) *For $v \in \Gamma$ and $n \gg 0$, we have the identity*

$$JS(v, n) = \sum_{\substack{v_1, \ldots, v_l \in \Gamma, \\ l \geq 1, v_1 + \cdots + v_l = v, \\ \overline{\chi}(v_i, m) = \overline{\chi}(v, m).}} \frac{(-1)^l}{l!} \prod_{i=1}^{l} (-1)^{\chi(\gamma_n - v_1 - \cdots - v_{i-1}, v_i)} \qquad (2.15)$$

$$\cdot \chi(\gamma_n - v_1 - \cdots - v_{i-1}, v_i) DT_H(v_i).$$

Here $\gamma_n := \mathrm{ch}(\mathcal{O}_X(-n))$ and $\chi(-, -)$ is the bilinear form (2.9).

Example

Suppose that $v_m = m \cdot \mathrm{ch}(\mathcal{O}_x)$ for a closed point $x \in X$. It is not easy to compute $DT_H(v_m)$ from its definition, and Theorem 2.3 gives a practical way to compute it. It is easy to see that the JS pair for $v = v_m$ and $n = 0$ is nothing but the pair (E, s) where E is a zero-dimensional sheaf with length m and $s: \mathcal{O}_X \to E$ is a surjection. Therefore $M_X^{JS}(v_m, 0)$ is isomorphic to the Hilbert scheme of m-points $\mathrm{Hilb}^m(X)$, or equivalently the moduli space $M_X^{H\text{-st}}(1, 0, 0, -m)$ (see Sect. 1.5). It follows that we have $JS(v_m, 0) = I_{m,0}$, whose generating series is given by (1.24).

In this case, we can apply Theorem 2.3 for $n = 0$, and the formula (2.3) gives the LHS of the identity (2.15). From this, one can compute $DT_H(v_m)$ as follows (see [92, Sect. 6.3]):

$$N_m := DT_H(v_m) = - \sum_{k \geq 1, k|m} \frac{e(X)}{k^2}. \qquad (2.16)$$

Another application of Theorem 2.3 is the deformation invariance of generalized DT invariants. Since JS pair invariants are defined via the virtual classes, their deformation invariance is a general consequence of the theory of virtual class. By the

comparison of JS pair invariants and generalized DT invariants in Theorem 2.3, we
have the following:

Theorem 2.4 ([92, Corollary 5.28]) *The deformation invariance of Theorem 1.3
holds for the generalized DT invariants* $\mathrm{DT}_{H_t}(v_t) \in \mathbb{Q}$ *(without assuming that*
$M_{\mathcal{X}_t}^{H_t\text{-st}}(v_t)$ *is fine, satisfies the ss=st condition).*

Chapter 3
Donaldson–Thomas Invariants for Quivers with Super-Potentials

Abstract The original DT invariants are defined for projective CY 3-folds. However, using the description of the DT invariants via Behrend functions, we can also define DT invariants for quivers with super-potentials. In this chapter, we review the theory of quivers with super-potentials and the associated DT invariants.

3.1 Introduction to Quivers with Super-Potentials

We first recall the notion of quivers, whose basic reference is [19, Sect. 4.1].

Definition 3.1 A *quiver* is a quadruple

$$Q = (Q_0, Q_1, s, t),$$

where Q_0 is a finite set of vertices, Q_1 is a finite set of edges, and $s, t\colon Q_1 \to Q_0$ are maps giving the source and the target of each edge. The *path algebra* $\mathbb{C}[Q]$ of a quiver Q is an associative algebra with basis all paths, where a path is a composition of edges:

$$\bullet^{a_1} \xrightarrow{e_1} \bullet^{a_2} \xrightarrow{e_2} \cdots \xrightarrow{e_{n-1}} \bullet^{a_n} \tag{3.1}$$

Here we also allow paths of length zero at each vertex. The product of two paths is their concatenation if they are composable, and zero otherwise. A *relation of a quiver* Q is a two-sided ideal $I \subset \mathbb{C}[Q]$.

Example

Let $Q_0 = \{1, 2\}$, $Q_1 = \{e_1, e_2\}$ such that $s(e_1) = s(e_2) = 1, t(e_1) = t(e_2) = 2$. This quiver is described by the following picture:

$$\bullet^1 \underset{e_2}{\overset{e_1}{\rightrightarrows}} \bullet^2$$

Y. Toda, *Recent Progress on the Donaldson–Thomas Theory*,
SpringerBriefs in Mathematical Physics 43,
https://doi.org/10.1007/978-981-16-7838-7_3

In this case, the path algebra is $\mathbb{C}[Q] = \begin{pmatrix} \mathbb{C} & \mathbb{C}^2 \\ 0 & \mathbb{C} \end{pmatrix}$.

We next recall the notion of super-potentials, whose basic references are [55, 69].

Definition 3.2 A *super-potential* of a quiver Q is an element

$$W \in \mathbb{C}[Q]/[\mathbb{C}[Q], \mathbb{C}[Q]].$$

Here $[-, -]$ is the commutator.

Remark 3.1 Note that any path p described as (3.1) is an element of $[\mathbb{C}[Q], \mathbb{C}[Q]]$ if $a_1 \neq a_n$. This is because $p = [e_{a_n}, p]$, where e_{a_n} is the length-zero path at a_n. Therefore any super-potential is written as a linear combination of cycles, i.e. a path (3.1) such that $a_1 = a_n$.

Let $p = e_1 e_2 \cdots e_n$ be a path as in (3.1). Then for an edge $e \in Q_1$, we set

$$p/e := \begin{cases} e_2 \cdots e_n, & \text{if } e = e_1 \\ 0, & \text{if } e \neq e_1. \end{cases}$$

Suppose that the path p above is a cycle, i.e. $a_1 = a_n$. Then the derivation of p by e is defined by

$$\frac{\partial p}{\partial e} := (e_1 e_2 \cdots e_n/e) + (e_2 e_3 \cdots e_1/e) + \cdots + (e_n e_1 \cdots e_{n-1}/e).$$

For a super-potential W, we define $\partial W/\partial e$ by the linearity, which is well-defined since $\partial p/\partial e = 0$ if $p \in [\mathbb{C}[Q], \mathbb{C}[Q]]$. Then we have the following relation of Q, defined to be the two-sided ideal generated by $\partial W/\partial e$ for all $e \in Q_1$:

$$\partial W := \left(\frac{\partial W}{\partial e} : e \in Q_1 \right) \subset \mathbb{C}[Q]. \tag{3.2}$$

Example

Let (Q, W) be a quiver with super-potential as in (3.3). Then $\mathbb{C}[Q]$ is the free algebra $\mathbb{C}\langle X, Y, Z \rangle$, and the two sided ideal (3.2) is generated by commutators $[X, Y], [Y, Z]$, $[Z, X]$. Therefore the quotient algebra $\mathbb{C}[Q]/(\partial W)$ is isomorphic to the polynomial ring $\mathbb{C}[X, Y, Z]$.

$$W = XYZ - XZY. \tag{3.3}$$

3.2 Moduli Spaces of Representations of Quivers

We recall the notion of representations of quivers as follows:

Definition 3.3 A *representation of a quiver* Q is a collection of vector spaces V_i for each $i \in Q_0$ and linear maps $f_e \colon V_{s(e)} \to V_{t(e)}$ for each $e \in Q_1$. If each V_i is finite dimensional, the collection of their dimensions

$$(\dim V_i)_{i \in Q_0} \in \mathbb{Z}^{Q_0}$$

is called the *dimension vector*.

The category of representations of Q is an abelian category, which we denote by $\mathrm{Rep}(Q)$. For an element $v = (v_i)_{i \in Q_0} \in \mathbb{Z}^{Q_0}$, let V_i be finite-dimensional vector spaces with $\dim V_i = v_i$. We consider the following quotient stack:

$$\mathcal{M}_Q(v) = \left[\bigoplus_{e \in Q_1} \mathrm{Hom}(V_{s(e)}, V_{t(e)}) / \prod_{i \in Q_0} \mathrm{GL}(V_i) \right].$$

Here $\prod_{i \in Q_0} \mathrm{GL}(V_i)$ acts on $\prod_{e \in Q_1} \mathrm{Hom}(V_{s(e)}, V_{t(e)})$ by

$$(g_i)_{i \in Q_0} \cdot (f_e)_{e \in Q_1} = (g_{t(e)} \circ f_e \circ g_{s(e)}^{-1})_{e \in Q_1}.$$

By the construction, the stack $\mathcal{M}_Q(v)$ is the moduli stack of Q-representations of dimension vector v. Its good moduli space is given by (see (1.5))

$$\pi_Q \colon \mathcal{M}_Q(v) \longrightarrow M_Q(v) := \bigoplus_{e \in Q_1} \mathrm{Hom}(V_{s(e)}, V_{t(e)}) /\!/ \prod_{i \in Q_0} \mathrm{GL}(V_i). \tag{3.4}$$

We then consider stability conditions on quiver representations following [101, Definition 1.1].

Definition 3.4 For a dimension vector $v = (v_i)_{i \in Q_0} \in \mathbb{Z}^{Q_0}$, let $\theta = (\theta_i)_{i \in Q_0} \in \mathbb{R}^{Q_0}$ satisfy that

$$\theta(v) := \sum_{i \in Q_0} \theta_i \cdot v_i = 0.$$

We say that a Q-representation (V_i, f_e) of dimension vector v is θ-*(semi)stable* if for any non-zero proper subrepresentation (V_i', f_e') of (V_i, f_e) of dimension vector v', we have $\theta(v') < (\leq)0$.

Similarly to (1.4), we can consider θ-(semi)stable loci in the moduli stack of representations and their good moduli spaces. They are summarized in the following diagram:

$$
\begin{array}{ccccc}
\mathscr{M}_Q^{\theta\text{-st}}(v) & \hookrightarrow & \mathscr{M}_Q^{\theta\text{-ss}}(v) & \hookrightarrow & \mathscr{M}_Q(v) \\
\downarrow & & \downarrow & & \downarrow \searrow^{\mathrm{tr}(W)} \\
M_Q^{\theta\text{-st}}(v) & \hookrightarrow & M_Q^{\theta\text{-ss}}(v) & \longrightarrow & M_Q(v) \xrightarrow[\overline{\mathrm{tr}}(W)]{} \mathbb{C}.
\end{array}
\tag{3.5}
$$

Here the arrows $\hookrightarrow$ are open immersions; vertical arrows are good moduli space morphisms. The left vertical arrow is a $\mathbb{C}^*$-gerbe and $M_Q^{\theta\text{-st}}(v)$ is smooth (see [101, Sect. 4]). The maps $\mathrm{tr}(W)$, $\overline{\mathrm{tr}}(W)$ will be given by a super-potential W as explained in the next section.

Remark 3.2 In [101, Proposition 3.1], it is proved that the θ-(semi)stable locus in $\oplus_{E \in Q_1} \mathrm{Hom}(V_{s(e)}, V_{t(E)})$ coincides with the GIT-(semi)stable locus with respect to some character of $\prod_{i \in Q_0} \mathrm{GL}(V_i)$. So the good moduli space $M_Q^{\theta\text{-ss}}(v)$ is obtained as a quasi-projective GIT quotient, which parametrizes θ-polystable Q-representations.

3.3 DT Invariants for Quivers with Super-Potentials

Let W be a super-potential of a quiver Q, given by a finite sum of cycles

$$W = \sum c_{e_\bullet} \cdot e_1 e_2 \cdots e_n,$$

where $e_1 e_2 \cdots e_n$ is a path as in (3.1) such that $a_1 = a_n$ and $c_{e_\bullet} \in \mathbb{C}$. Then we have the map $\mathrm{tr}(W) \colon \mathscr{M}_Q(v) \to \mathbb{A}^1$ defined by

$$\mathrm{tr}(W)(f_e)_{e \in Q_1} = \sum c_{e_\bullet} \cdot \mathrm{tr}(f_{e_n} \circ \cdots \circ f_{e_1}). \tag{3.6}$$

Here $f_{e_n} \circ \cdots \circ f_{e_1} \in \mathrm{End}(V_{a_1})$ and we take its trace. The above function is invariant under the $\prod_{i \in Q_0} \mathrm{GL}(V_i)$-action, so determines a function on $\mathscr{M}_Q(v)$. The critical locus of $\mathrm{tr}(W)$

$$\mathscr{M}_{(Q,W)}(v) := \mathrm{Crit}(\mathrm{tr}(W)) \subset \mathscr{M}_Q(v) \tag{3.7}$$

corresponds to (Q, W)-representations, i.e. Q-representations satisfying the relation (3.2). Similarly to (3.5), we have the θ-(semi)stable loci inside the above stack, and their good moduli spaces:

$$
\begin{array}{ccccc}
\mathscr{M}^{\theta\text{-st}}_{(Q,W)}(v) & \hookrightarrow & \mathscr{M}^{\theta\text{-ss}}_{(Q,W)}(v) & \hookrightarrow & \mathscr{M}_{(Q,W)}(v) \\
\downarrow & & \downarrow & & \downarrow \\
M^{\theta\text{-st}}_{(Q,W)}(v) & \hookrightarrow & M^{\theta\text{-ss}}_{(Q,W)}(v) & \longrightarrow & M_{(Q,W)}(v).
\end{array}
$$

The function $\mathrm{tw}(W)$ on $\mathscr{M}_Q(v)$ factors through the map from the good moduli space $M_Q(v)$ by Definition 1.3 (ii), giving the function $\overline{\mathrm{tr}}(W)$ in the diagram (3.5). Since the left arrow in (3.5) is a $\mathbb{C}^*$-gerbe, we have

$$
M^{\theta\text{-st}}_{(Q,W)}(v) = \mathrm{Crit}\left(\overline{\mathrm{tr}}(W)|_{M^{\theta\text{-st}}_Q(v)}\right) \subset M^{\theta\text{-st}}_Q(v),
$$

i.e. the moduli space of θ-stable (Q, W)-representation $M^{\theta\text{-st}}_{(Q,W)}(v)$ is a global critical locus inside the smooth ambient space $M^{\theta\text{-st}}_Q(v)$. Therefore it is reasonable to consider its Behrend function, and the DT invariants for (Q, W) is defined as an analogy of the identity (1.21) as follows:

Definition 3.5 Suppose that $\mathscr{M}^{\theta\text{-st}}_{(Q,W)}(v) = \mathscr{M}^{\theta\text{-ss}}_{(Q,W)}(v)$ holds. In this case, we define $\mathrm{DT}_\theta(v) \in \mathbb{Z}$ by

$$
\mathrm{DT}_\theta(v) := \int_{M^{\theta\text{-st}}_{(Q,W)}(v)} \chi_B \, de.
$$

Here χ_B is the Behrend constructible function on $M^{\theta\text{-st}}_{(Q,W)}(v)$.

Remark 3.3 Similarly to Definition 2.5, we can also define generalized DT invariant $\mathrm{DT}_\theta(v) \in \mathbb{Q}$ without the assumption $\mathscr{M}^{\theta\text{-st}}_{(Q,W)}(v) = \mathscr{M}^{\theta\text{-ss}}_{(Q,W)}(v)$, using the motivic Hall algebra of (Q, W)-representations. See [92, Definition 7.15].

Example

Let (Q, W) be a quiver with a super-potential given as follows:

$$W = X^{n+1}.$$

Let us take the dimension vector $v \in \mathbb{Z}$, and take $\theta = 0$. Then we have $\mathscr{M}_Q^{\theta\text{-st}}(v) = \mathscr{M}_Q^{\theta\text{-ss}}(v)$ if and only if $v = 1$. For $v = 1$, we have

$$M_Q^{\theta\text{-st}}(1) = \mathrm{Spec}\mathbb{C}[x]/x^n,$$

which is the critical locus of $f : \mathbb{C} \to \mathbb{C}$, $f(x) = x^{n+1}$. By computing the Behrend function, we have $\mathrm{DT}_\theta(1) = (-1)^1(1 - (n+1)) = n$.

Example

Let (Q, W) be a quiver with a super-potential given as follows

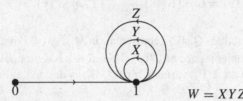

$$W = XYZ - XZY.$$

Let $v = (v_0, v_1) = (1, n) \in \mathbb{Z}^2$ be the dimension vector, and set $\theta = (\theta_0, \theta_1)$ satisfying $\theta_0 + n\theta_1 = 0$. By taking $\theta_1 < 0$, a θ-stable Q-representation of dimension vector v can be shown to be isomorphic to the Hilbert scheme of n-points in $\mathbb{C}^3$ and satisfies $\mathscr{M}_Q^{\theta\text{-st}}(v) = \mathscr{M}_Q^{\theta\text{-ss}}(v)$. Then we have

$$\mathrm{DT}_\theta(1, n) = \int_{\mathrm{Hilb}^n(\mathbb{C}^3)} \chi_B \, de.$$

Their generating series is computed in [15, Theorem 4.11] (compare with the formula (1.24)):

$$\sum_{n \geq 0} \mathrm{DT}_\theta(1, n)q^n = M(-q).$$

3.4 Non-commutative DT Invariants on Resolved Conifold

Let Q be a quiver

$$\tag{3.8}$$

with super-potential $W = a_1 b_1 a_2 b_2 - a_1 b_2 a_2 b_1$. Note that Q contains a full sub-quiver Q_0 consisting of vertices $\{0, 1\}$. A (Q, W)-representation (V_∞, V_0, V_1) is called *cyclic* if the image of $V_\infty \to V_0$ generates $V_0 \oplus V_1$ as $\mathbb{C}[Q_0]$-module. Let us take a dimension vector $v = (v_\infty, v_0, v_1) = (1, m_0, m_1)$ and a stability condition $\theta = (\theta_0, \theta_0, \theta_1)$ satisfying $\theta(v) = 0$. The following lemma is easy to prove (see [135, 158]).

Lemma 3.1 *For a generic choice of* $\theta = (\theta_0, \theta_1)$, *we have* $\mathcal{M}^{\theta\text{-st}}_{(Q,W)}(v) = \mathcal{M}^{\theta\text{-ss}}_{(Q,W)}(v)$. *Moreover, if* $\theta_0 < 0$, $\theta_1 < 0$, *the above moduli space consists of cyclic* (Q, W)-*representations.*

For a generic θ, we set the following generating series:

$$A_\theta(Q, W) = \sum_{m_0 \geq 0, m_1 \geq 0} \mathrm{DT}_\theta(1, m_0, m_1) q_0^{m_0} q_1^{m_1}.$$

If $\theta_0 < 0, \theta_1 < 0$, the coefficients of the above generating series count cyclic (Q, W)-representations by Lemma 3.1. The following result was conjectured in [158] and proved in [135, 192].

Theorem 3.1 ([135, 158, 192]) *For* $\theta_0 < 0, \theta_1 < 0$, *we have the identity*

$$A_\theta(Q, W) = M(-q_0 q_1)^2 \prod_{k \geq 1} \left(1 + q_0^k (-q_1)^{k-1}\right)^k \left(1 + q_0^k (-q_1)^{k+1}\right)^k.$$

Remark 3.4 The above quiver with super-potential is related to the non-compact CY 3-fold

$$X = \mathrm{Tot}(\mathscr{O}_{\mathbb{P}^1}(-1)^{\oplus 2}) \longrightarrow \mathbb{P}^1,$$

which is a resolution of the conifold singularity $\{xy - zw = 0\} \subset \mathbb{C}^4$. Let $\mathscr{P} = \mathscr{O}_X \oplus \mathscr{O}_X(1)$. Van den Bergh [191] showed that there exists an equivalence

$$\Phi := \mathbf{RHom}(\mathscr{P}, -) \colon D^b(X) \xrightarrow{\sim} D^b(A),$$

Fig. 3.1 Wall–chamber
structures

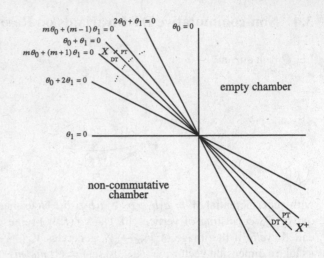

where A is the non-commutative algebra $\text{End}(\mathscr{P})$, which is shown to be isomorphic to $\mathbb{C}[Q]/(\partial W)$ for the above quiver with super-potential (Q, W). See [135, 158] for the relation of the formula in Theorem 3.1, the DT (or PT) invariants on X and its flop X^+.

The generating series $A_\theta(Q, W)$ depends on θ, and Nagao–Nakajima [135] proved Theorem 3.1 by investigating the wall-crossing formula for the θ-stability. The wall–chamber structure obtained in [135] is described in Fig. 3.1. For example, suppose that θ lies in the following wall for $m \geq 1$:

$$\theta \in \{(\theta_0, \theta_1) \in \mathbb{R}^2 : m\theta_0 + (m - 1)\theta_1 = 0, \theta_0 < 0\}.$$

Then for $\theta_\pm = \theta \mp (\varepsilon, 0)$ for $0 < \varepsilon \ll 1$, the wall-crossing formula is given by (see [135, Theorem 3.12])

$$A_{\theta_+}(Q, W) = \left(1 + q_0^m(-q_1)^{m+1}\right)^m \cdot A_{\theta_-}(Q, W).$$

On the other hand, it is easy to see that the moduli space is empty for $\theta_0 > 0, \theta_1 > 0$, so $A_\theta(Q, W) = 1$ in such a chamber. By applying the above wall-crossing formula from the empty chamber to the non-commutative chamber in Fig. 3.1, the formula in Theorem 3.1 is obtained.

3.5 Analytic Neighborhood Theorem

Finally in this chapter, we explain a relationship between moduli spaces of semistable sheaves on CY 3-folds and those of representations of quivers with super-potentials proved in [183]. Roughly speaking, the latter is interpreted as a local theory of the former on the good moduli spaces.

We first prepare analytic versions of super-potentials. For a quiver Q, let $\mathbb{C}[\![Q]\!]$ be its *completed path algebra*. An element of $\mathbb{C}[\![Q]\!]$ is a formal infinite sum

$$f = \sum c_{e_\bullet} \cdot e_1 e_2 \cdots e_n \tag{3.9}$$

for $c_{e_\bullet} \in \mathbb{C}$, where $e_1 e_2 \cdots e_n$ is a path in Q.

Definition 3.6 ([183, Definition 2.1, 2.17])

(1) The subalgebra $\mathbb{C}\{Q\} \subset \mathbb{C}[\![Q]\!]$ is defined to be elements (3.9) such that $|c_{e_\bullet}| < C^n$ for some constant $C > 0$ which is independent of n.

(2) A *convergent super-potential* is an element

$$W \in \mathbb{C}\{Q\}/\overline{[\mathbb{C}\{Q\}, \mathbb{C}\{Q\}]}.$$

Here $\overline{(-)}$ is the topological closure of $(-)$. .

For a convergent super-potential W, the trace function (3.6) is an infinite sum, but has a convergence radius by the convergence condition in Definition 3.6 (1). Indeed, the trace function (3.6) converges to give an analytic function $\mathrm{tr}(W)_U$ on $\pi_Q^{-1}(U)$ for some analytic open neighborhood $0 \in U \subset M_Q(v)$, where π_Q is the good moduli space morphism (3.4) (see [183, Sect. 2.6]). By taking the critical locus, we obtain the moduli stack of (Q, W)-representations (analytic locally on the good moduli space)

$$\mathscr{M}_{(Q,W)}(v)_U := \mathrm{Crit}(\mathrm{tr}(W)_U) \subset \pi_Q^{-1}(U).$$

We next define the Ext-quiver associated with a collection of sheaves.

Definition 3.7 Let $E_\bullet = (E_1, E_2, \ldots, E_k)$ be a collection of coherent sheaves on a projective variety X. The *Ext-quiver* $Q_{E_\bullet}$ associated with $E_\bullet$ has vertices $\{1, 2, \ldots, k\}$, and the number of edges from i to j is $\dim \mathrm{Ext}^1(E_i, E_j)$.

Let X be a smooth projective CY 3-fold, H an ample divisor and take $\gamma \in \Gamma$. Recall that the moduli stack of H-semistable sheaves $\mathscr{M}_X^{H\text{-ss}}(\gamma)$ admits a good moduli space (see the diagram (1.6))

$$\pi_X^H : \mathscr{M}_X^{H\text{-ss}}(\gamma) \longrightarrow M_X^{H\text{-ss}}(\gamma),$$

where the latter space parametrizes H-polystable sheaves of the form

$$E = \bigoplus_{i=1}^{k} E_i^{\oplus v_i}, \quad \gamma = \sum_{i=1}^{k} v_i \cdot \mathrm{ch}(E_i). \tag{3.10}$$

Here $(E_1, E_2, \ldots, E_k)$ is a collection of mutually non-isomorphic H-stable sheaves.

Theorem 3.2 ([183, Corollary 5.7]) *For any point $p \in M_X^{H\text{-ss}}(\gamma)$ corresponding to a H-polystable sheaf (3.10), let $Q_{E_\bullet}$ be the Ext-quiver associated with the collection $(E_1, E_2, \ldots, E_k)$. Let $v = (v_i)_{1 \le i \le k}$ be the dimension vector of $Q_{E_\bullet}$, determined by v_i in (3.10). Then there exist analytic open subsets $p \in V \subset M_X^{H\text{-ss}}(\gamma)$, $0 \in U \subset M_{Q_{E_\bullet}}(v)$ and a convergent super-potential $W_{E_\bullet}$ of $Q_{E_\bullet}$ such that there exist an isomorphism of analytic stacks*

$$\mathscr{M}_{(Q_{E_\bullet}, W_{E_\bullet})}(v)_U \xrightarrow{\cong} (\pi_X^H)^{-1}(V).$$

Remark 3.5 The super-potential $W_{E_\bullet}$ is constructed from minimal A_∞-structure on the dg-category generated by $(E_1, E_2, \ldots, E_k)$. See [183, Sect. 5.5] for details.

Remark 3.6 The result of Theorem 3.2 is useful to prove some properties of DT invariants which are already known to those for quivers with super-potentials. See [165] for an application to wall-crossing of Gopakumar–Vafa invariants, using the Davison–Meinhardt PBW theorem for cohomological DT invariants for quivers with super-potentials (see Theorem 6.3). Also, since a version of Conjecture 2.1 for a quiver with super-potential follows from [52, Theorem A] (see Remark 6.10), the result of Theorem 3.2 may be used to prove Conjecture 2.1.

Chapter 4
Donaldson–Thomas Invariants for Bridgeland Semistable Objects

Abstract In Chaps. 1, 2, the DT invariants are defined as counting (semi)stable sheaves on CY 3-folds. In this chapter, we extend them to invariants counting (semi)stable objects in the derived category of coherent sheaves. For this purpose, we need a suitable notion of stability in the derived category. The notion of stability conditions on derived categories (more generally triangulated categories) was introduced by Bridgeland [27] as a mathematical framework of Douglas's Π-stability [58] in physics. In this chapter, we give an introduction to Bridgeland stability conditions [27], and introduce DT-type invariants counting Bridgeland semistable objects.

4.1 Definition of Bridgeland Stability Conditions

Recall the notion of (semi)stable sheaves discussed in Sect. 1.1. In the case of curves, we observed in Remark 1.2 that semistable sheaves are characterized by the inequality of slopes among all the subsheaves. This is rephrased in the following way. Suppose that C is a smooth projective curve and Z is a group homomorphism

$$Z \colon K(C) \longrightarrow \mathbb{C}, \ E \longmapsto -\deg(E) + i\operatorname{rank}(E). \qquad (4.1)$$

It has the following property:

(i) For any non-zero $E \in \operatorname{Coh}(C)$, we have

$$Z(E) \in \mathbb{H} := \{z \in \mathbb{C} : \operatorname{Im} z > 0 \text{ or } z \in \mathbb{R}_{<0}\}.$$

An object $E \in \operatorname{Coh}(C)$ is (semi)stable if and only if for any non-zero subobject $E' \subsetneq E$, we have $\arg Z(E') < (\leq) \arg Z(E)$ in $(0, \pi]$.

(ii) For any $E \in \operatorname{Coh}(C)$, there exists an HN filtration.

The idea of Bridgeland stability conditions is to generalize the above two conditions for an arbitrary heart of the bounded t-structure of a triangulated category.

© The Author(s), under exclusive license to Springer Nature Singapore Pte Ltd. 2021 39
Y. Toda, *Recent Progress on the Donaldson—Thomas Theory*,
SpringerBriefs in Mathematical Physics 43,
https://doi.org/10.1007/978-981-16-7838-7_4

Let $\mathscr{D}$ be a triangulated category (see [76] for generalities on triangulated categories) and [1] the shift functor on it, e.g. $\mathscr{D} = D^b(X)$ is the bounded derived category of coherent sheaves on a variety X. Recall that a subcategory $\mathscr{A} \subset \mathscr{D}$ is called the *heart of a bounded t-structure* if the following conditions are satisfied (see [27, Lemma 3.2]):

(i) For any $E_1, E_2 \in \mathscr{A}$ and $i < 0$, we have $\mathrm{Hom}(E_1, E_2[i]) = 0$.

(ii) For any $E \in \mathscr{A}$, there exists a finite sequence of distinguished triangles

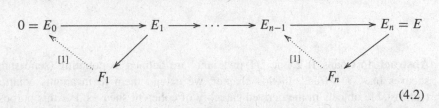

$$(4.2)$$

such that $F_i \in \mathscr{A}[k_i]$ for some $k_i \in \mathbb{Z}$ with $k_1 > k_2 > \cdots > k_n$. For each $k \in \mathbb{Z}$, we set $\mathscr{H}_{\mathscr{A}}^k(F) := F_i$ if $k = -k_i$ for some i, and $\mathscr{H}_{\mathscr{A}}^k(F) = 0$ otherwise.

For example, the subcategory $\mathrm{Coh}(X) \subset D^b(X)$ consisting of complexes located in degree zero is the heart of a bounded t-structure such that $\mathscr{H}_{\mathrm{Coh}(X)}^k(-)$ is the usual cohomology sheaves. It is proved in [17] that the heart of a bounded t-structure $\mathscr{A} \subset \mathscr{D}$ is an abelian category. Here is a definition of Bridgeland stability conditions.

Definition 4.1 ([27, Definition 5.1, Proposition 5.3]) For a triangulated category $\mathscr{D}$, a *Bridgeland stability condition* on it is a pair $(Z, \mathscr{A})$, where $Z \colon K(\mathscr{D}) \to \mathbb{C}$ is a group homomorphism (called *central charge*) and $\mathscr{A} \subset \mathscr{D}$ is the heart of a bounded t-structure satisfying the following conditions:

(i) For any non-zero $E \in \mathscr{A}$, we have $Z(E) \in \mathbb{H}$. An object $E \in \mathscr{A}$ is called *Z-(semi)stable* if for any non-zero subobject $E' \subsetneq E$, we have the inequality $\arg Z(E') < (\leq) \arg Z(E)$.

(ii) For any non-zero $E \in \mathscr{A}$, there exists a filtration (called *HN filtration*)

$$0 = E_0 \subset E_1 \subset \cdots \subset E_n = E \tag{4.3}$$

such that each subquotient $F_i = E_i/E_{i-1}$ is Z-semistable with $\arg Z(F_1) > \arg Z(F_2) > \cdots > \arg Z(F_n)$.

It is easy to see that an HN filtration is unique up to isomorphism if it exists.

Example

Let C be a smooth projective curve and $\mathscr{D} = D^b(C)$. Then $(Z, \mathrm{Coh}(C))$ is a Bridgeland stability condition, where Z is given by (4.1). An object $E \in \mathrm{Coh}(C)$ is Z-(semi)stable if and only if it is a H-(semi)stalbe sheaf for any polarization H on C.

Example

Let $Q = (Q_0, Q_1, s, t)$ be a finite quiver, and $\mathrm{Rep}(Q)$ the abelian category of finite dimensional Q-representations. For each choice of $\theta_i \in \mathbb{H}$ for $i \in Q_0$, let Z be the group homomorphism

$$K(\mathrm{Rep}(Q)) \xrightarrow{\dim} \bigoplus_{i \in Q_0} \mathbb{Z}e_i \longrightarrow \mathbb{C},$$

where the first arrow is taking the dimension vector and the second arrow takes e_i to θ_i. Then the pair $(Z, \mathrm{Rep}(Q))$ is a Bridgeland stability condition on $D^b(\mathrm{Rep}(Q))$.

4.2 The Space of Bridgeland Stability Conditions

A fundamental result on Bridgeland stability conditions is that the set of stability conditions satisfying some nice properties forms a complex manifold. Let Γ be a finitely generated free abelian group with a norm $\|-\|$ on $\Gamma_{\mathbb{R}}$. Let $\mathscr{D}$ be a triangulated category and suppose that we have a surjective group homomorphism

$$\mathrm{cl} \colon K(\mathscr{D}) \longrightarrow \Gamma.$$

For example, if $\mathscr{D} = D^b(X)$ for a smooth projective variety X, we often take Γ to be the image of the Chern character map as in (1.2), and take $\mathrm{cl} = \mathrm{ch}$.

Definition 4.2 We define $\mathrm{Stab}_\Gamma(\mathscr{D})$ to be the set of Bridgeland stability conditions $(Z, \mathscr{A})$ satisfying the following:

(i) (*numerical condition*) there is a group homomorphism $Z' \colon \Gamma \to \mathbb{C}$ such that we have the factorization as $Z \colon K(\mathscr{D}) \xrightarrow{\mathrm{cl}} \Gamma \xrightarrow{Z'} \mathbb{C}$,
(ii) (*support property*) we have

$$\sup \left\{ \frac{\|\mathrm{cl}(E)\|}{|Z(E)|} : E \text{ is } Z\text{-semistable} \right\} < \infty.$$

Note that Z' is uniquely determined since $\mathrm{cl} \colon K(\mathscr{D}) \to \Gamma$ is surjective. The following is the main result in [27].

Theorem 4.1 ([27, Theorem 1.2]) *There is a natural topology on $\mathrm{Stab}_\Gamma(\mathscr{D})$ such that the forgetting map*

$$\mathrm{Stab}_\Gamma(\mathscr{D}) \longrightarrow \Gamma_{\mathbb{C}}^{\vee}, \ (Z, \mathscr{A}) \longmapsto Z'$$

is a local homeomorphism. In particular, $\mathrm{Stab}_\Gamma(\mathscr{D})$ is a complex manifold whose complex structure is induced from that of $\Gamma_{\mathbb{C}}^{\vee}$.

Remark 4.1 The support property was not imposed in Bridgeland's original paper [27]. If there is no support property, the result of Theorem 4.1 only holds for some linear subspace in $\Gamma_{\mathbb{C}}^{\vee}$, as formulated in [27, Theorem 1.2]. The support property first appeared in Kontsevich–Soibelman's paper [108], which turned out to be useful to prove several foundational properties of the space of stability conditions, e.g. local deformation property, existence of a wall–chamber structure, etc.

Here we explain the topology on $\mathrm{Stab}_\Gamma(\mathscr{D})$. Given an element $\sigma = (Z, \mathscr{A}) \in \mathrm{Stab}_\Gamma(\mathscr{D})$ and $\phi \in (0, 1]$ we define

$$\mathscr{P}(\phi) := \{E \in \mathscr{A} : E \text{ is } Z\text{-semistable with } Z(E) \in \mathbb{R}_{>0}e^{i\pi\phi}\} \cup \{0\}.$$

For an arbitrary $\phi \in \mathbb{R}$, we define $\mathscr{P}(\phi) \subset \mathscr{D}$ by the rule $\mathscr{P}(\phi + 1) = \mathscr{P}(\phi)[1]$. A non-zero object in $\mathscr{P}(\phi)$ is called a σ-*semistable object with phase* ϕ. By combining the sequence of distinguished triangles (4.4) and HN filtrations (4.3), we have the following: for any $E \in \mathscr{D}$ there exists a sequence of distinguished triangles

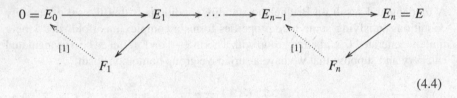

$$(4.4)$$

such that $F_i \in \mathscr{P}(\phi_i)$ with $\phi_1 > \cdots > \phi_n$, which is unique up to isomorphisms. We set $\phi_\sigma^+(E) := \phi_1$ and $\phi_\sigma^-(E) := \phi_n$. For $\sigma, \tau \in \mathrm{Stab}_\Gamma(\mathscr{D})$, we set

$$d(\sigma, \tau) := \sup_{0 \neq E \in \mathscr{D}} \{|\phi_\sigma^+(E) - \phi_\tau^+(E)|, |\phi_\sigma^-(E) - \phi_\tau^-(E)|\}.$$

For each $\sigma = (Z, \mathscr{A}) \in \mathrm{Stab}_\Gamma(\mathscr{D})$ and $\varepsilon_1 > 0, \varepsilon_2 > 0$, we define

$$B_{\varepsilon_1, \varepsilon_2}(\sigma) := \{\tau = (W, \mathscr{B}) \in \mathrm{Stab}_\Gamma(\mathscr{D}) : \|Z - W\| < \varepsilon_1, d(\sigma, \tau) < \varepsilon_2\}.$$

Then the topology on $\mathrm{Stab}_\Gamma(\mathscr{D})$ is given so that the subsets $B_{\varepsilon_1, \varepsilon_2}(\sigma)$ for $\sigma \in \mathrm{Stab}_\Gamma(\mathscr{D})$, $\varepsilon_1 > 0, \varepsilon_2 > 0$ form an open basis (see [27, Sect. 6]).

Definition 4.3 For a smooth projective variety X and $\mathrm{cl} = \mathrm{ch} \colon K(X) \to \Gamma$ as in (1.2), we define the *space of Bridgeland stability conditions* by

$$\mathrm{Stab}(X) := \mathrm{Stab}_\Gamma(D^b(X)).$$

Example

Let C be an elliptic curve. Then $\mathrm{Stab}(C)$ is isomorphic to $\widetilde{\mathrm{GL}}_2^+(\mathbb{R})$ (see [27, Theorem 9.1]). Here $\widetilde{\mathrm{GL}}_2^+(\mathbb{R})$ is the universal covering space of the subgroup $\mathrm{GL}_2^+(\mathbb{R}) \subset \mathrm{GL}_2(\mathbb{R})$ preserving the orientation on $\mathbb{R}^2$.

4.3 Constructions of Bridgeland Stability Conditions

One of the fundamental problems in the theory of stability conditions is that, in general, it is not known whether $\mathrm{Stab}(X)$ is non-empty or not. This is obvious in the dim $X = 1$ case as we discussed in Sect. 4.1. However, in the dim $X \geq 2$ case, there is no stability condition $(Z, \mathscr{A})$ such that $\mathscr{A} = \mathrm{Coh}(X)$ is the standard heart (see [168, Lemma 2.7]), so we need to construct some other heart $\mathscr{A}$ to find an element of $\mathrm{Stab}(X)$. Let $A(X)_{\mathbb{C}}$ be the *complexified ample cone* of X

$$A(X)_{\mathbb{C}} := \{B + iH \in H^2(X, \mathbb{C}) : H \text{ is ample}\}.$$

For each $B + iH \in A(X)_{\mathbb{C}}$, from the literature in physics [58] it is expected that the group homomorphism

$$Z_{B,H} \colon K(X) \longrightarrow \mathbb{C}, \ E \longmapsto -\int_X e^{-iH} \mathrm{ch}^B(E) \tag{4.5}$$

together with some heart $\mathscr{A} \subset D^b(X)$ gives a stability condition $(Z_{B,H}, \mathscr{A}) \in \mathrm{Stab}(X)$. Here $\mathrm{ch}^B(E) := e^{-B}\mathrm{ch}(E)$.

The following notion plays a key role to construct other hearts.

Definition 4.4 ([75]) Let $\mathscr{A}$ be an abelian category. A pair of full subcategories $(\mathscr{T}, \mathscr{F})$ on $\mathscr{A}$ is called a *torsion pair* if the following conditions are satisfied:

 (i) For any $T \in \mathscr{T}$ and $F \in \mathscr{F}$, we have $\mathrm{Hom}(T, F) = 0$.
 (ii) For any $E \in \mathscr{A}$, there exists an exact sequence $0 \to T \to E \to F \to 0$ in $\mathscr{A}$ such that $T \in \mathscr{T}$ and $F \in \mathscr{F}$.

Example

For a variety X, let $\mathscr{T} \subset \mathrm{Coh}(X)$ be the subcategory consisting of torsion sheaves and $\mathscr{F} \subset \mathrm{Coh}(X)$ consisting of torsion-free sheaves. Then $(\mathscr{T}, \mathscr{F})$ is a torsion pair on $\mathrm{Coh}(X)$.

Given a torsion pair in the heart of a bounded t-structure, we can produce another heart called tilting.

Definition 4.5 ([75]) Let $\mathscr{D}$ be a triangulated category and $\mathscr{A} \subset \mathscr{D}$ the heart of a bounded t-structure. Then for a torsion pair $(\mathscr{T}, \mathscr{F})$ on $\mathscr{A}$, its *tilting* $\mathscr{A}^\dagger$ is defined by

$$\mathscr{A}^\dagger := \{E \in \mathscr{D} : \mathscr{H}_{\mathscr{A}}^0(E) \in \mathscr{T}, \mathscr{H}_{\mathscr{A}}^{-1}(E) \in \mathscr{F}, \mathscr{H}_{\mathscr{A}}^i(E) = 0 \text{ for } i \neq -1, 0\}.$$

As proved in [75], the heart $\mathscr{A}^\dagger \subset \mathscr{D}$ is also the heart of a bounded t-structure. We denote the tilting as $\mathscr{A}^\dagger = \langle \mathscr{F}[1], \mathscr{T} \rangle$.

For $B + iH \in H^2(X, \mathbb{C})$ as above and $E \in \mathrm{Coh}(X)$, let $\mu_{B,H}(E)$ be defined by

$$\mu_{B,H}(E) := \frac{H^{d-1} \cdot \mathrm{ch}_1^B(E)}{\mathrm{ch}_0^B(E)} \in \mathbb{R} \cup \{\infty\}.$$

Here $d = \dim X$ and $\mu_{B,H}(E) = \infty$ when the denominator is zero (we also set $\mu_{B,H}(E) = \infty$ when both of the denominator and the numerator are zero). The above slope function defines the following $\mu_{B,H}$-stability.

Definition 4.6 An object $E \in \mathrm{Coh}(X)$ is called $\mu_{B,H}$-*(semi)stable* if for any non-zero subsheaf $E' \subsetneq E$, we have $\mu_{B,H}(E') < (\leq)\mu_{B,H}(E/E')$.

Remark 4.2 The inequality $\mu_{B,H}(E') \leq \mu_{B,H}(E/E')$ can be replaced by $\mu_{B,H}(E') \leq \mu_{B,H}(E)$. However, if $\dim X \geq 2$, the inequality $\mu_{B,H}(E') < \mu_{B,H}(E/E')$ cannot be replaced by $\mu_{B,H}(E') < \mu_{B,H}(E)$ as otherwise there are no stable torsion free sheaves. The same applies to Definition 4.7.

Remark 4.3 For $B = 0$, the above $\mu_{B,H}$-stability is related to the H-stability in Definition 1.1 in the following way:

$$\mu_{0,H}\text{-stable} \implies H\text{-stable} \implies H\text{-semistable} \implies \mu_{0,H}\text{-semistable}.$$

Remark 4.4 Using the slope function (4.6), one may try to construct a stability condition of the form $(Z, \mathrm{Coh}(X))$, where Z is

$$Z(E) = -H^{d-1} \cdot \mathrm{ch}_1^B(E) + i\mathrm{ch}_0^B(E).$$

However, the above pair $(Z, \mathrm{Coh}(X))$ does not determine a Bridgeland stability condition if $\dim X \geq 2$, since $Z(\mathscr{O}_x) = 0$ for each closed point $x \in X$.

Similarly to the H-stability on $\mathrm{Coh}(X)$, for any $E \in \mathrm{Coh}(X)$ there exists an HN filtration

$$0 = E_0 \subset E_1 \subset \cdots E_n = E$$

such that each $F_i = E_i/E_{i-1}$ is $\mu_{B,H}$-semistable and $\mu_{B,H}(F_1) > \cdots > \mu_{B,H}(F_n)$. We set $\mu_{B,H}^+(E) := \mu_{B,H}(F_1)$, $\mu_{B,H}^-(E) := \mu_{B,H}(F_n)$. Then from the existence of HN filtration, the subcategories

$$\mathcal{T}_{B,H} := \{E \in \mathrm{Coh}(X) : \mu_{B,H}^-(E) > 0\}, \quad \mathcal{F}_{B,H} := \{E \in \mathrm{Coh}(X) : \mu_{B,H}^+(E) \le 0\}$$

form a torsion pair of $\mathrm{Coh}(X)$. By taking the tilting, we obtain another heart

$$\mathcal{B}_{B,H} = \langle \mathcal{F}_{B,H}[1], \mathcal{T}_{B,H} \rangle \subset D^b(X).$$

Remark 4.5 A striking feature of this heart is that for any object $E \in \mathcal{B}_{B,H}$ we have $H^{d-1}\mathrm{ch}_1^B(E) \ge 0$, i.e. $H^{d-1}\mathrm{ch}_1^B(-)$ behaves like a rank of the usual coherent sheaves.

Now let us consider the surface case, i.e. $\dim X = 2$. In this case, the group homomorphism $Z_{B,H}$ in (4.5) is written as

$$Z_{B,H}(E) = -\mathrm{ch}_2^B(E) + \frac{1}{2}H^2\mathrm{ch}_0^B(E) + iH\mathrm{ch}_1^B(E).$$

From Remark 4.5, we have $\mathrm{Im} Z_{B,H}(E) \ge 0$ for any $E \in \mathcal{B}_{B,H}$, which is a necessary condition for the pair $(Z_{B,H}, \mathcal{B}_{B,H})$ to give a Bridgeland stability condition. Indeed, the above pair is known to give a Bridgeland stability condition.

Proposition 1 ([5, 28, 176]) *For* $\dim X = 2$, *we have* $(Z_{B,H}, \mathcal{B}_{B,H}) \in \mathrm{Stab}(X)$.

In order to prove the above proposition, one has to show that if $\mathrm{Im} Z_{B,\omega}(E) = 0$, then $\mathrm{Re} Z_{B,\omega}(E) < 0$. This is less obvious part, and guaranteed by the following *Bogomolov–Gieseker (BG) inequality* (see [5, Corollary 2.1] for how to use the BG inequality to derive Proposition 1):

Theorem 4.2 ([21, 68]) *For any torsion-free* $\mu_{B,H}$-*semistable sheaf* E, *we have the inequality*

$$H^{d-2}(\mathrm{ch}_1^B(E)^2 - 2\mathrm{ch}_0^B(E)\mathrm{ch}_2^B(E)) \ge 0.$$

4.4 Bogomolov–Gieseker-type Inequality

Let us consider the case of $\dim X = 3$. In this case, the group homomorphism $Z_{B,H}$ in (4.5) is written as

$$Z_{B,H}(E) = -\mathrm{ch}_3^B(E) + \frac{1}{2}H^2\mathrm{ch}_1^B(E) + i\left(H\mathrm{ch}_2^B(E) - \frac{1}{6}H^3\mathrm{ch}_0^B(E)\right).$$

Since we cannot control the sign of $\mathrm{Im} Z_{B,H}(E)$ for $E \in \mathcal{B}_{B,H}$, the pair $(Z_{B,H}, \mathcal{B}_{B,H})$ does not give a Bridgeland stability condition.

An idea in the paper [8] is to construct a desired heart by a further tilting of $\mathcal{B}_{B,H}$. A key observation in [8, Lemma 3.2.1] is that the vector

$$(H^2\mathrm{ch}_1^B(E), \mathrm{Im}Z_{B,H}(E), -\mathrm{Re}Z_{B,H}(E))$$

on $\mathcal{B}_{B,H}$ behaves like (rank, ch_1, ch_2) on surfaces, i.e. for non-zero $E \in \mathcal{B}_{B,H}$ we have:

(i) $H^2\mathrm{ch}_1^B(E) \geq 0$,
(ii) if $H^2\mathrm{ch}_1^B(E) = 0$ then $\mathrm{Im}Z_{B,H}(E) \geq 0$,
(iii) if $H^2\mathrm{ch}_1^B(E) = \mathrm{Im}Z_{B,H}(E) = 0$, then $-\mathrm{Re}Z_{B,H}(E) > 0$.

Therefore it is natural to consider the following slope function on $\mathcal{B}_{B,H}$:

$$\nu_{B,H}(E) = \frac{\mathrm{Im}Z_{B,H}(E)}{H^2\mathrm{ch}_1^B(E)} \in \mathbb{R} \cup \{\infty\}.$$

Then we have the following analogue of slope stability, called *tilt stability* in [8].

Definition 4.7 ([8, Definition 3.2.3]) A non-zero object $E \in \mathcal{B}_{B,H}$ is called $\nu_{B,H}$-*(semi)stable* if for any non-zero subobject $E' \subsetneq E$, we have the inequality $\nu_{B,H}(E') < (\leq)\nu_{B,H}(E/E')$.

Below we take B, H to be rational classes. Then similarly to the $\mu_{B,H}$-stability on $\mathrm{Coh}(X)$, the $\nu_{B,H}$-stability admits HN filtrations in $\mathcal{B}_{B,H}$. Therefore similarly to $(\mathcal{T}_{B,H}, \mathcal{F}_{B,H})$ on $\mathrm{Coh}(X)$, we have the torsion pair in $\mathcal{B}_{B,H}$,

$$\mathcal{T}'_{B,H} := \{E \in \mathcal{B}_{B,H} : \nu_{B,H}^-(E) > 0\}, \quad \mathcal{F}'_{B,H} := \{E \in \mathcal{B}_{B,H} : \nu_{B,H}^+(E) \leq 0\}.$$

We have the associated tilting

$$\mathcal{A}_{B,H} = \langle \mathcal{F}'_{B,H}[1], \mathcal{T}'_{B,H} \rangle \subset D^b(X).$$

The following is the main conjecture in [8].

Conjecture 4.1 ([8, Conjecture 3.2.6]) If $\dim X = 3$, we have $(Z_{B,H}, \mathcal{A}_{B,H}) \in \mathrm{Stab}(X)$.

Recall that the Bogomolov–Gieseker inequality in Theorem 4.2 is required to construct Bridgeland stability conditions on surfaces. In the 3-fold case, since $Z_{B,H}$ involves the third Chern character, one may think that Bogomolov–Gieseker-type inequality evaluating the third Chern character may be required for Conjecture 4.1. Such an inequality is proposed in [8, Conjecture 1.3.1], and reformulated in [7, Conjecture 4.1].

Definition 4.8 We say that a pair $(X, B + iH)$, where X is a smooth projective 3-fold and $B + iH \in A(X)_{\mathbb{C}}$, satisfies the *BG-type inequality* if for any non-zero $\nu_{B,H}$-semistable object $E \in \mathcal{B}_{B,H}$, we have the inequality

$$(H^2\mathrm{ch}_1^B(E))^2 - 2H^3\mathrm{ch}_0^B(E) \cdot H\mathrm{ch}_2^B(E)$$
$$+ 12(H\mathrm{ch}_2^B(E))^2 - 18H^2\mathrm{ch}_1^B(E) \cdot \mathrm{ch}_3^B(E) \geq 0. \qquad (4.6)$$

If $(X, B + iH)$ satisfies the BG-type inequality, then for any $\nu_{B,H}$-semistable object $E \in \mathscr{B}_{B,H}$ with $\nu_{B,H}(E) = 0$, an easy calculation shows that

$$\mathrm{ch}_3^B(E) \leq \frac{1}{18} H^2 \mathrm{ch}_1^B(E). \tag{4.7}$$

Indeed, the inequality (4.7) for any tilt-semistable object with $\nu_{B,H}(-) = 0$ is equivalent to the inequality (4.6) for any tilt-semistable object (see [7, Theorem 4.2]). The inequality (4.7) implies that for any $0 \neq E \in \mathscr{A}_{B,H}$ we have $Z_{B,\omega}(E) \in \mathbb{H}$ (see [8, Sect. 3.2]). Indeed, when $\rho(X) = 1$, the BG-type inequality is enough to construct Bridgeland stability conditions. Here $\rho(X) := \dim \mathrm{NS}(X)_{\mathbb{R}}$ is the *Picard number of* X.

Proposition 2 ([7, Theorem 8.2]) *Suppose that* $(X, B + iH)$ *satisfies the BG-type inequality and* X *satisfies* $\rho(X) = 1$. *Then Conjecture 4.1 holds.*

Remark 4.6 Suppose that $(X, B + iH)$ satisfies the BG-type inequality but $\rho(X) > 1$. Then the pair $(Z_{B,H}, \mathscr{A}_{B,H})$ satisfies the axiom of Definition 4.1, but the inequality (4.6) is not strong enough to prove the support property. See [137, Theorem 3.20] for the proof of the support property when X is an abelian 3-fold with $\rho(X) > 1$, using Fourier–Mukai transforms.

The BG-type inequality is expected to hold if every effective divisor D on X is nef, i.e. $D \cdot C \geq 0$ for any curve $C \subset X$, e.g. when $\rho(X) = 1$.

Conjecture 4.2 ([8, Conjecture 1.3.1], [7, Conjecture 4.1]) *Let* X *be a smooth projective 3-fold such that any effective divisor on it is nef. Then any pair* $(X, B + iH)$ *satisfies the BG-type inequality.*

Remark 4.7 It was conjectured in [8, Conjecture 1.3.1] that the inequality (4.7) holds in general. However, it turned out later that the inequality (4.7) is not true in general, e.g. a blow-up at a point $X \to \mathbb{P}^3$ (see [154]). Note that the exceptional divisor is not nef in this case. If there is a non-nef effective divisor, it seems that we need to add some modification term of the BG-type inequality. See [20] in the case of Fano 3-folds.

Remark 4.8 At this moment, Conjecture 4.2 is known to hold in the following cases: X is a Fano 3-fold with $\rho(X) = 1$ ([8, 115]), X is an abelian 3-fold ([7, 123]), X has a nef tangent bundle ([110]), and X is a quintic 3-fold ([114]). Also see [121] for another approach to construct stability conditions on product varieties.

The most important result among those in Remark 4.8 is Chunyi Li's work to settle Conjecture 4.2 for quintic 3-folds with some (B, H).

Theorem 4.3 ([114, Theorem 2.8]) *Let* $X \subset \mathbb{P}^4$ *be a smooth quintic hypersurface, which is a CY 3-fold. Then* $(X, B + iH)$ *satisfies the BG-type inequality if* $B = \beta \cdot c_1(\mathscr{O}_X(1))$, $H = \sqrt{3}\alpha \cdot c_1(\mathscr{O}_X(1))$ *for* $(\alpha, \beta) \in \mathbb{R}^2$ *satisfying the following:*

$$\alpha^2 + \left(\beta - [\beta] - \frac{1}{2}\right)^2 > \frac{1}{4}.$$

Remark 4.9 In the context of mirror symmetry, the space of stability conditions is expected to be related to the stringy Kähler moduli space, which is the moduli space of complex structures on a mirror manifold (see [29, Sect. 7]). In the case of quintic 3-folds, it is expected that we have the embedding

$$[\{\psi \in \mathbb{C} : \psi^5 \neq 1\}/\mu_5] \longrightarrow [\text{Auteq}(X)\backslash\text{Stab}(X)/\mathbb{C}]. \tag{4.8}$$

The LHS is the stringy Kähler moduli space of X. The above embedding should be described in terms of solutions of Picard–Fuchs equations and is highly transcendental (see [181, Sect. 3]).

There are three special points on a compactification of the LHS of (4.8): $\psi = \infty$ (*large volume limit*), $\psi = 1$ (*conifold point*) and $\psi = 0$ (*Gepner point*). The stability conditions in Theorem 4.3 should form a neighborhood at the large volume limit. The Gepner point and a neighborhood at the conifold point have not yet been constructed.

4.5 Moduli Spaces of Bridgeland Semistable Objects

Let X be a smooth projective variety with a Bridgeland stability condition $\sigma = (Z, \mathcal{A}) \in \text{Stab}(X)$. Then we can consider a moduli problem of σ-semistable objects in $D^b(X)$. We first consider moduli theory of objects in $D^b(X)$ without stability conditions. For $v \in \Gamma$, we define the 2-functor

$$\widetilde{\mathcal{M}}_X(v) \colon (Sch/\mathbb{C}) \longrightarrow (Groupoid)$$

sending a $\mathbb{C}$-scheme T to the groupoid of objects $\mathcal{E} \in D^b(X \times T)$ such that for any closed point $t \in T$, the derived pull-back $\mathcal{E}_t := \mathbf{L}i_t^*\mathcal{E}$ for $i_t \colon X \times \{t\} \hookrightarrow X \times T$ is an object in $D^b(X)$ satisfying $\text{ch}(\mathcal{E}_t) = v$ and

$$\text{Ext}^{<0}(\mathcal{E}_t, \mathcal{E}_t) = 0. \tag{4.9}$$

By the result of Lieblich [120], the 2-functor $\widetilde{\mathcal{M}}_X(v)$ is an Artin stack locally of finite type.

Remark 4.10 The condition (4.9) is necessary for the 2-functor $\widetilde{\mathcal{M}}_X(v)$ to be an Artin stack. Otherwise we have to consider the ∞-functor where the target is the ∞-category of simplical sets, and the relevant moduli problem is represented by an Artin ∞-stack (see [188]).

The condition (4.9) is automatically satisfied if $\mathcal{E}_t$ lies in a heart of a bounded t-structure on $D^b(X)$. In particular, the condition (4.9) holds for any Bridgeland semistable object. Therefore we have the substacks

$$\mathscr{M}_X^{\sigma\text{-st}}(v) \subset \mathscr{M}_X^{\sigma\text{-ss}}(v) \subset \widetilde{\mathscr{M}}_X(v)$$

consisting of σ-(semi)stable objects in the heart $\mathscr{A}$. Similarly to the case of (semi)stable sheaves (1.4), it is natural to expect the following:

Conjecture 4.3 ([150, Conjecture 1.1]) The moduli stacks $\mathscr{M}_X^{\sigma\text{-st(ss)}}(v)$ are Artin stacks of finite type, and are open substacks of $\widetilde{\mathscr{M}}_X(v)$.

Suppose that $\mathscr{M}_X^{\sigma\text{-ss}}(v)$ is an Artin stack of finite type. Then by [3, Theorem 7.25], we have the following diagram similar to (1.6):

$$
\begin{array}{ccc}
\mathscr{M}_X^{\sigma\text{-st}}(v) & \hookrightarrow & \mathscr{M}_X^{\sigma\text{-ss}}(v) \\
\downarrow & \square & \downarrow \\
M_X^{\sigma\text{-st}}(v) & \hookrightarrow & M_X^{\sigma\text{-ss}}(v).
\end{array}
\tag{4.10}
$$

Here the horizontal arrows are open immersions, the vertical arrows are good moduli space morphisms, such that $M_X^{\sigma\text{-ss}}(v)$ is a proper algebraic space and the left arrow is a $\mathbb{C}^*$-gerbe. In particular, if $\mathscr{M}_X^{\sigma\text{-st}}(v) = \mathscr{M}_X^{\sigma\text{-ss}}(v)$, then $M_X^{\sigma\text{-st}}(v)$ is a proper algebraic space.

Remark 4.11 Conjecture 4.3 is known to be true in many cases, but in general this is an open question. In the K3 surface case, the above question is proved in [167, Theorem 4.12], and the same argument applies to any surface for stability conditions in Proposition 1.

Let X be a smooth projective 3-fold. Below we assume that $\rho(X) = 1$ for simplicity. Then if a pair $(X, B + iH)$ for $B + iH \in A(X)_{\mathbb{C}}$ satisfies the BG-type inequality in Definition 4.8, we have an element $(Z_{B,H}, \mathscr{A}_{B,H}) \in \text{Stab}(X)$ by Proposition 2. We denote by $\text{Stab}^\circ(X) \subset \text{Stab}(X)$ the connected component which contains the above stability conditions. The following result is proved in [150].

Theorem 4.4 ([150, Theorem 4.2]) *Let X be a smooth projective 3-fold with $\rho(X) = 1$ satisfying Conjecture 4.2. Then Conjecture 4.3 is true for any $\sigma \in \text{Stab}^\circ(X)$.*

Remark 4.12 In [150, Theorem 4.2], the above result is proved for a subset of $\text{Stab}^\circ(X)$, denoted by $\text{Stab}^\circ_{\omega,B}(X)$, without the assumption of the Picard rank. The $\rho(X) = 1$ assumption implies that $\text{Stab}^\circ_{\omega,B}(X) = \text{Stab}^\circ(X)$, so [150, Theorem 4.2] implies Theorem 4.4.

The moduli stacks in the above theorem are generalizations of classical semistable sheaves in (1.4) by the following proposition (see [122] for a similar result for surfaces).

Proposition 3 ([137, Proposition 3.26]) *Let X be a smooth projective 3-fold with $\rho(X) = 1$ satisfying Conjecture 4.2, and H an ample divisor on X. Then for any $v \in \Gamma$, there exists $\sigma \in \text{Stab}^\circ(X)$ such that $\mathscr{M}_X^{\sigma\text{-st(ss)}}(v) = \mathscr{M}_X^{H\text{-st(ss)}}(v)$.*

Note that the moduli stacks $\mathcal{M}_X^{\sigma\text{-st(ss)}}(v)$ depend on a choice of σ, and if we take σ different from the one in Proposition 3 we may obtain new moduli stacks beyond classical moduli stacks.

From a general theory of Bridgeland stability conditions, there exists a *wall–chamber structure* on $\text{Stab}^\circ(X)$, i.e. for each $v \in \Gamma$ there exist locally finite real codimension one submanifolds

$$\bigcup_{i \in I} \mathcal{W}_i \subset \text{Stab}^\circ(X)$$

called *walls* satisfying the following: for each connected component (called a *chamber*)

$$\mathcal{C} \subset \text{Stab}^\circ(X) \setminus \bigcup_{i \in I} \mathcal{W}_i$$

the moduli stacks $\mathcal{M}_X^{\sigma\text{-st(ss)}}(v)$ are constant for $\sigma \in \mathcal{C}$, while they may change when σ crosses a wall (see [28, Proposition 9.3]).

4.6 DT-type Invariants for Bridgeland Semistable Objects

Let X be a smooth projective CY 3-fold with $\rho(X) = 1$ satisfying Conjecture 4.2. Then by Theorem 4.4, for $\sigma \in \text{Stab}^\circ(X)$ we have moduli stacks of σ-(semi)stable objects $\mathcal{M}_X^{\sigma\text{-st(ss)}}(v)$ which are Artin stacks of finite type. If $\mathcal{M}_X^{\sigma\text{-st}}(v) = \mathcal{M}_X^{\sigma\text{-ss}}(v)$, then the good moduli space $M_X^{\sigma\text{-st}}(v)$ is a proper algebraic space (see the diagram (4.10)). Furthermore, if it admits a universal object, there is a zero-dimensional virtual fundamental class on $M_X^{\sigma\text{-st}}(v)$ and its integration gives the DT invariant counting σ-stable objects (see [150, Definition 5.2]). In general, $\mathcal{M}_X^{\sigma\text{-st}}(v) \neq \mathcal{M}_X^{\sigma\text{-ss}}(v)$, and in such a case the generalized DT invariant is defined using motivic Hall algebras as in Definition 2.5 (see [150, Definition 5.5]).

More precisely, the above DT invariants for Bridgeland semistable objects are defined as follows. For $\sigma = (Z, \mathcal{A}) \in \text{Stab}^\circ(X)$, let $\mathcal{M}_\mathcal{A}$ be the stack of objects in $\mathcal{A}$. By using $\mathcal{M}_\mathcal{A}$, we can define the motivic Hall algebra $H(\mathcal{A})$ of $\mathcal{A}$ by (see Definition 2.2)

$$H(\mathcal{A}) := K(\text{St}/\mathcal{M}_\mathcal{A}).$$

Then similarly to $H(\text{Coh}(X))$, we have the $*$-product on $H(\mathcal{A})$, the subalgebra of regular elements $H^{\text{reg}}(\mathcal{A}) \subset H(\mathcal{A})$, the Poisson algebra $H^{\text{sc}}(\mathcal{A})$ and the Poisson algebra homomorphism $I: H^{\text{sc}}(\mathcal{A}) \to C(X)$ (see Sect. 2.3). Moreover, we have elements (see Sect. 2.4)

$$\delta^{\sigma\text{-ss}}(v) := [\mathcal{M}_X^{\sigma\text{-ss}}(v) \subset \mathcal{M}_{\mathscr{A}}] \in H(\mathscr{A}),$$

$$\epsilon^{\sigma\text{-ss}}(v) := \sum_{k \geq 1} \sum_{\substack{v_1+\cdots+v_k=v \\ \arg Z(v_i)=\arg Z(v)}} \frac{(-1)^{k-1}}{k} \delta^{\sigma\text{-ss}}(v_1) * \cdots * \delta^{\sigma\text{-ss}}(v_k) \in H(\mathscr{A}).$$

Then similarly to Theorem 2.2, we have

$$\bar{\epsilon}^{\sigma\text{B}}(v) := (\mathbb{L}-1) \cdot \epsilon^{\sigma\text{-ss}}(v) \in H^{\text{reg}}(\mathscr{A}), \tag{4.11}$$

and define $\widehat{\epsilon}^{\sigma\text{-ss}}(v)$ to be the above class in $H^{\text{sc}}(\mathscr{A})$. The invariant $\mathrm{DT}_\sigma(v)$ is defined by (see Definition 2.5)

$$I(\widehat{\epsilon}^{\sigma\text{-ss}}(v)) = -\mathrm{DT}_\sigma(v) \cdot c_v.$$

As a summary, we have the following:

Theorem 4.5 ([150, Theorem 5.6], [137, Proposition 4.2]) *In the above setting, there exists a map*

$$\mathrm{DT}_*(v): \mathrm{Stab}^\circ(X) \longrightarrow \mathbb{Q}, \ \sigma \longmapsto \mathrm{DT}_\sigma(v), \tag{4.12}$$

where $\mathrm{DT}_\sigma(v)$ virtually counts σ-semistable objects with Chern character v. Moreover, for an ample divisor H on X, there exists $\sigma \in \mathrm{Stab}^\circ(X)$ such that $\mathrm{DT}_H(v) = \mathrm{DT}_\sigma(v)$ for some $\sigma \in \mathrm{Stab}^\circ(X)$, where $\mathrm{DT}_H(v)$ is defined in Definition 2.5.

Chapter 5
Wall-Crossing Formulas of Donaldson–Thomas Invariants

Abstract In this chapter, we explain wall-crossing formulas of DT invariants and their applications. The applications include DT/PT correspondence, rationality of PT generating series, flop formula of DT invariants, etc.

5.1 An Idea of Wall-Crossing and Applications

Let X be a smooth projective CY 3-fold with $\rho(X) = 1$, and consider DT invariants counting Bridgeland semistable objects in Theorem 4.5. Recall that by the wall-chamber structure on $\mathrm{Stab}^\circ(X)$, the invariant $\mathrm{DT}_\sigma(v)$ is constant if σ lies on a chamber but may change when σ crosses a wall. So the following problem is natural to address:

Problem 5.1 How do the invariants $\mathrm{DT}_\sigma(v) \in \mathbb{Q}$ change when σ crosses a wall?

As we will see below, there is a wall-crossing formula [92, 108] which answers the above problem.

Before discussing the wall-crossing formula, we explain one of its possible applications. Let Y be a smooth projective CY 3-fold with an equivalence of derived categories

$$\Phi \colon D^b(X) \overset{\sim}{\longrightarrow} D^b(Y).$$

For example, we have such a derived equivalence if Y is birational to X by Bridgeland [26]. Let us consider the problem of comparing classical DT invariants on X and Y counting stable sheaves, e.g. rank one DT invariants counting curves in X and Y. Since Φ is an equivalence, it induces the isomorphism of the spaces of stability conditions

$$\Phi_* \colon \mathrm{Stab}(X) \overset{\cong}{\longrightarrow} \mathrm{Stab}(Y). \tag{5.1}$$

© The Author(s), under exclusive license to Springer Nature Singapore Pte Ltd. 2021
Y. Toda, *Recent Progress on the Donaldson—Thomas Theory*,
SpringerBriefs in Mathematical Physics 43,
https://doi.org/10.1007/978-981-16-7838-7_5

Let Γ_X be the image of the Chern character map as in (1.2). Then the equivalence Φ induces the isomorphism $\phi \colon \Gamma_X \overset{\cong}{\to} \Gamma_Y$. By Theorem 4.5, for each $v \in \Gamma_X$ and ample divisors H_X, H_Y on X, Y, there exist $\sigma_X \in \mathrm{Stab}^\circ(X), \sigma_Y \in \mathrm{Stab}^\circ(Y)$ such that we have

$$\mathrm{DT}_{H_X}(v) = \mathrm{DT}_{\sigma_X}(v), \ \mathrm{DT}_{H_Y}(v') = \mathrm{DT}_{\sigma_Y}(v') = \mathrm{DT}_{\sigma'_Y}(v),$$

where $v' = \phi(v)$ and $\sigma'_Y = \Phi_*^{-1}(\sigma_Y)$. Here the last identity is tautological from the derived equivalence Φ. Therefore the problem comparing $\mathrm{DT}_{H_X}(v)$ and $\mathrm{DT}_{H_Y}(v')$ is reduced to comparing $\mathrm{DT}_{\sigma_X}(v), \mathrm{DT}_{\sigma'_Y}(v)$.

Suppose furthermore that the isomorphism (5.1) preserves the connected components $\mathrm{Stab}^\circ(X)$ and $\mathrm{Stab}^\circ(Y)$. Then there is a path which connects σ_X and σ'_Y in $\mathrm{Stab}^\circ(X)$. By applying the wall-crossing formula at each wall on the above path, we obtain a relation between $\mathrm{DT}_{\sigma_X}(v)$ and $\mathrm{DT}_{\sigma'_Y}(v)$, so a relation between $\mathrm{DT}_{H_X}(v)$ and $\mathrm{DT}_{H_Y}(v')$. In particular, if $X = Y$, then Φ is an autoequivalence of $D^b(X)$ and we obtain a constraint of $\mathrm{DT}_{H_X}(v)$ induced by Φ.

For example, the above idea is used in [170] to prove the rationality of PT generating series (see Sect. 5.4), and in [38, 175] to prove a flop transformation formula of curve-counting DT invariants (see Sect. 5.5).

5.2 General Principle of Wall-Crossing Formulas

In this section, we explain a general idea of wall-crossing formulas in the setting of Theorem 4.5. Let X be a smooth projective CY 3-fold with $\rho(X) = 1$ satisfying Conjecture 4.2. We take $v \in \Gamma$ and $\sigma = (Z, \mathcal{A}) \in \mathrm{Stab}^\circ(X)$ such that $\arg Z(v) \in (0, \pi)$. Suppose that σ lies in a wall with respect to v, and take $\tau = (W, \mathcal{B}) \in \mathrm{Stab}^\circ(X)$ which is sufficiently close to σ and lies in an adjacent chamber. Then for each σ-semistable object $E \in \mathcal{A}$ with $\mathrm{ch}(E) = v$, there exists an HN filtration in $\mathcal{A}$ with respect to τ

$$0 = E_0 \subset E_1 \subset \cdots \subset E_k = E \tag{5.2}$$

such that each $F_i = E_i/E_{i-1}$ is τ-semistable satisfying

$$\arg W(F_1) > \cdots > \arg W(F_k), \ \arg Z(F_1) = \cdots = \arg Z(F_k).$$

Here $\arg Z(F_i) \leq \arg Z(E)$ as E is Z-semistable and $\arg W(F_i) > \arg W(E)$, so we have $\arg Z(F_i) = \arg Z(E)$ since we took τ to be sufficiently close to σ (see the proof of [89, Theorem 5.11]). An idea to deduce the wall-crossing formula is to describe the above HN filtration in terms of motivic Hall algebras. From the filtration (5.2) and the definition of motivic Hall algebra, one can show that the following identity holds in $H(\mathcal{A})$ (see [89, Theorem 5.11]):

$$\delta^{\sigma\text{-ss}}(v) = \sum_{\substack{k \geq 1}} \sum_{\substack{v_1 + \cdots + v_k = v \\ \arg W(v_1) > \cdots > \arg W(v_k) \\ \arg Z(v_1) = \cdots = \arg Z(v_k)}} \delta^{\tau\text{-ss}}(v_1) * \cdots * \delta^{\tau\text{-ss}}(v_k). \tag{5.3}$$

We first discuss the simplest case of wall-crossing. Suppose that $v \in \Gamma$ is primitive, and also satisfies the assumption:

$$\max\{k : v_1 + \cdots + v_k = v, \ Z(v_i) \in \mathbb{R}_{>0} Z(v)\} = 2. \tag{5.4}$$

The above assumption easily implies that $k \leq 2$ in the formula (5.3). Moreover, for each v_i in (5.3), we have the following identities:

$$\epsilon^{\sigma\text{-ss}}(v_i) = \delta^{\sigma\text{-ss}}(v_i), \ \epsilon^{\tau\text{-ss}}(v_i) = \delta^{\tau\text{-ss}}(v_i).$$

Here we have assumed $v_i \neq v$ for the first identity. The second identity holds for $v_i = v$ as v is primitive and τ lies in a chamber. Therefore the formula (5.3) is simplified as

$$\delta^{\sigma\text{-ss}}(v) = \epsilon^{\tau\text{-ss}}(v) + \sum_{\substack{v_1 + v_2 = v \\ \arg W(v_1) > \arg W(v_2) \\ \arg Z(v_1) = \arg Z(v_2)}} \epsilon^{\tau\text{-ss}}(v_1) * \epsilon^{\tau\text{-ss}}(v_2).$$

On the other hand, from the definition of $\epsilon^{\sigma\text{-ss}}(v)$ and the assumption (5.4), we have the identity

$$\epsilon^{\sigma\text{-ss}}(v) = \delta^{\sigma\text{-ss}}(v) - \frac{1}{2} \sum_{\substack{v_1 + v_2 = v \\ \arg Z(v_1) = \arg Z(v_2)}} \delta^{\sigma\text{-ss}}(v_1) * \delta^{\sigma\text{-ss}}(v_2).$$

By the above two identities, an easy computation shows that

$$\epsilon^{\sigma\text{-ss}}(v) = \epsilon^{\tau\text{-ss}}(v) + \sum_{\substack{v_1 + v_2 = v \\ \arg W(v_1) > \arg W(v_2) \\ \arg Z(v_1) = \arg Z(v_2)}} \frac{1}{2} \left[\epsilon^{\tau\text{-ss}}(v_1), \epsilon^{\tau\text{-ss}}(v_2) \right].$$

By multiplying $(\mathbb{L} - 1)$ and taking the projection to $H^{\text{sc}}(\mathscr{A})$, we obtain the identity

$$\widehat{\epsilon}^{\sigma\text{-ss}}(v) = \widehat{\epsilon}^{\tau\text{-ss}}(v) + \sum_{\substack{v_1 + v_2 = v \\ \arg W(v_1) > \arg W(v_2) \\ \arg Z(v_1) = \arg Z(v_2)}} \frac{1}{2} \left\{ \widehat{\epsilon}^{\tau\text{-ss}}(v_1), \widehat{\epsilon}^{\tau\text{-ss}}(v_2) \right\}. \tag{5.5}$$

Then applying the Poisson algebra homomorphism in Theorem 2.1, we obtain the formula

$$\mathrm{DT}_\sigma(v) = \mathrm{DT}_\tau(v) \tag{5.6}$$

$$- \sum_{\substack{v_1+v_2=v \\ \arg W(v_1) > \arg W(v_2) \\ \arg Z(v_1) = \arg Z(v_2)}} \frac{(-1)^{\chi(v_1,v_2)}}{2} \chi(v_1,v_2)\mathrm{DT}_\tau(v_1)\mathrm{DT}_\tau(v_2).$$

Now suppose that $\tau' = (W', \mathscr{B}')$ is another stability condition contained in a chamber and sufficiently close to σ. Then the above formula also holds for τ'. Since σ does not lie on a wall for v_i by the condition (5.4), we have $\mathrm{DT}_\tau(v_i) = \mathrm{DT}_{\tau'}(v_i)$. Therefore we obtain the following wall-crossing formula:

$$\mathrm{DT}_{\tau'}(v) = \mathrm{DT}_\tau(v) + \sum_{\substack{v_1+v_2=v \\ \arg W'(v_1) > \arg W'(v_2) \\ \arg Z(v_1) = \arg Z(v_2)}} \frac{(-1)^{\chi(v_1,v_2)}}{2} \chi(v_1,v_2)\mathrm{DT}_\tau(v_1)\mathrm{DT}_\tau(v_2)$$

$$- \sum_{\substack{v_1+v_2=v \\ \arg W(v_1) > \arg W(v_2) \\ \arg Z(v_1) = \arg Z(v_2)}} \frac{(-1)^{\chi(v_1,v_2)}}{2} \chi(v_1,v_2)\mathrm{DT}_\tau(v_1)\mathrm{DT}_\tau(v_2).$$

Remark 5.1 In the above argument, we applied the Poisson algebra homomorphism in Theorem 2.1 to the identity (5.5) to derive (5.6). At least for a simpler naive Euler characteristic version as in Remark 2.2, the application of Poisson algebra homomorphism is much simpler under the assumption (5.4), and we do not need the full machinery of Theorem 2.1. Let $M_X^{\tau\text{-st}}(v_i)$ be the good moduli space of τ-stable objects (see the diagram (4.10)). From the assumption (5.4) (and also assuming that $M_X^{\tau\text{-st}}(v_i)$ is fine), one can check that

$$\widehat{\epsilon}^{\tau\text{-ss}}(v_i) = [M_X^{\tau\text{-st}}(v_i) \longrightarrow \mathcal{M}_{\mathscr{A}}].$$

So $\widehat{\epsilon}^{\tau\text{-ss}}(v_i)$ are delta-functions supported on τ-stable objects, and a computation similar to (2.11) is enough.

A general wall-crossing formula without the condition (5.4) follows from a similar argument, but is more complicated to describe. Let $l \subset \mathbb{H}$ be a semi-line which contains the origin as an endpoint. We define

$$\Gamma_{\sigma,l} := \{\mathrm{ch}(E) : E \text{ is } Z\text{-semistable and } Z(E) \in l\}.$$

We also set the following completion of the motivic Hall algebra:

$$\widehat{H}_{\sigma,l}(\mathscr{A}) := \prod_{v \in \Gamma_{\sigma,l}} H_v(\mathscr{A}).$$

We define the following elements of $\widehat{H}_{\sigma,l}(\mathscr{A})$:

$$\delta^{\sigma\text{-ss}}(l) := 1 + \sum_{v \in \Gamma_{\sigma,l}} \delta^{\sigma\text{-ss}}(v), \quad \epsilon^{\sigma\text{-ss}}(l) := \sum_{v \in \Gamma_{\sigma,l}} \epsilon^{\sigma\text{-ss}}(v).$$

Then as in Remark 2.3, they satisfy the following relation in $\widehat{H}_{\sigma,l}(\mathscr{A})$:

$$\epsilon^{\sigma\text{-ss}}(l) = \log(\delta^{\sigma\text{-ss}}(l)), \quad \delta^{\sigma\text{-ss}}(l) = \exp(\epsilon^{\sigma\text{-ss}}(l)). \tag{5.7}$$

Moreover, let $l' \subset \mathbb{H}$ be another semi-line which contains the origin as an endpoint, and set

$$\delta^{\tau\text{-ss}}(l') := 1 + \sum_{\substack{v \in \Gamma_{\sigma,l} \\ W(v) \in l'}} \delta^{\tau\text{-ss}}(v), \quad \epsilon^{\tau\text{-ss}}(l') := \sum_{\substack{v \in \Gamma_{\sigma,l} \\ W(v) \in l'}} \epsilon^{\tau\text{-ss}}(v).$$

Then they are also related by logarithm and the exponential as in the Eq. (5.7). Using the above identities, the identity (5.3) is alternatively written as

$$\delta^{\sigma\text{-ss}}(l) = \prod_{l'}^{\curvearrowright} \delta^{\tau\text{-ss}}(l').$$

Here the right hand side is the products of $\delta^{\tau\text{-ss}}(l')$ for all semi-lines as above in the clockwise order. Then using the relations (5.7), we obtain the following relation of ϵ-functions:

$$\epsilon^{\sigma\text{-ss}}(l) = \log\left(\prod_{l'}^{\curvearrowright} \exp(\epsilon^{\tau\text{-ss}}(l'))\right).$$

The point is that we can expand the above formula using Theorem 5.1 below, and express the relation of ϵ-functions in terms of Lie brackets: for each $v_1, \ldots, v_k \in \Gamma_{\sigma,l}$, there exists $U_{\sigma,\tau}(v_1, \ldots, v_k) \in \mathbb{Q}$ such that we have the identity in $H(\mathscr{A})$:

$$\epsilon^{\sigma\text{-ss}}(v)$$
$$= \sum_{k \geq 1} \sum_{v_1 + \cdots + v_k = v} U_{\sigma,\tau}(v_1, \ldots, v_k)[\epsilon^{\tau\text{-ss}}(v_1), [\epsilon^{\tau\text{-ss}}(v_2), \cdots [\epsilon^{\tau\text{-ss}}(v_{k-1}), \epsilon^{\tau\text{-ss}}(v_k)] \cdots]].$$

Here $[-, -]$ is the commutator, and $v, v_i \in \Gamma_{\sigma,l}$. We multiply both side by $(\mathbb{L} - 1)$ and take the projection onto $H^{\text{sc}}(\mathscr{A})$. By noting the identity

$$(\mathbb{L} - 1) \cdot [\epsilon^{\tau\text{-ss}}(v_1), [\epsilon^{\tau\text{-ss}}(v_2), \cdots [\epsilon^{\tau\text{-ss}}(v_{k-1}), \epsilon^{\tau\text{-ss}}(v_k)] \cdots]]$$
$$= \frac{1}{\mathbb{L} - 1}\left[\overline{\epsilon}^{\tau\text{-ss}}(v_1), \frac{1}{\mathbb{L} - 1}\left[\overline{\epsilon}^{\tau\text{-ss}}(v_2), \cdots \frac{1}{\mathbb{L} - 1}\left[\overline{\epsilon}^{\tau\text{-ss}}(v_{k-1}), \overline{\epsilon}^{\tau\text{-ss}}(v_k)\right] \cdots \right]\right],$$

where $\overline{\epsilon}^{\tau\text{-ss}}(v_i)$ is given by (4.11), we obtain the identity in $H^{\text{sc}}(\mathscr{A})$:

$$\widehat{\epsilon}_\sigma(v) = \sum_{k \geq 1} \sum_{v_1 + \cdots + v_k = v} U(v_1, \ldots, v_k) \{\widehat{\epsilon}_\tau(v_1), \{\widehat{\epsilon}_\tau(v_2), \cdots \{\widehat{\epsilon}_\tau(v_{k-1}), \widehat{\epsilon}_\tau(v_k)\} \cdots \}\}.$$

By applying the Poisson algebra homomorphism in Theorem 2.1, we obtain the formula

$$DT_\sigma(v) = \sum_{k \geq 1} \sum_{v_1 + \cdots + v_i = v} U_{\sigma,\tau}(v_1, \cdots, v_k)$$

$$\cdot (-1)^{\sum_{i<j} \chi(v_i, v_j) + k - 1} \prod_{i<j} \chi(v_i, v_j) \prod_{i=1}^{k} DT_\tau(v_i).$$

The above formula gives a relationship between DT invariants for σ and those for τ. Since it is of the form $DT_\sigma(v) = DT_\tau(v) + \cdots$, by the induction on $|Z(v)|$ for $v \in \Gamma_{\sigma,l}$ we can also invert the above equation so that $DT_\tau(v)$ can be also described in terms of $DT_\sigma(v_i)$. Then if τ' is another stability condition sufficiently close to σ but contained in another chamber, by combining the above arguments one can in principle describe $DT_{\tau'}(v)$ in terms of $DT_\tau(v)$.

In·the above argument, the combinatorial coefficients $U_{\sigma,\tau}(v_1, \cdots, v_k)$ are complicated to calculate, but in principle can be obtained using the following *Baker–Campbell–Hausdorff (BCH) formula* (see [89, Definition 4.4] for their explicit descriptions).

Theorem 5.1 (BCH formula) *Let* $\mathbb{C}\langle\!\langle X, Y \rangle\!\rangle$ *be the non-commutative formal power series ring. For* $X, Y \in \mathbb{C}\langle\!\langle X, Y \rangle\!\rangle$, *we have the identity*

$$\log(\exp(X) \cdot \exp(Y))$$

$$= \sum_{n>0} \frac{(-1)^{n-1}}{n} \sum_{\substack{r_i + s_i > 0 \\ 1 \leq i \leq n}} \frac{\left(\sum_{i=1}^{n}(r_i + s_i)\right)^{-1}}{r_1! s_1! \cdots r_n! s_n!}$$

$$\underbrace{[X, [X, \cdots [X,}_{r_1} \underbrace{[Y, [Y, \cdots, [Y,}_{s_1} \cdots \underbrace{[X, [X, \cdots [X,}_{r_n} \underbrace{[Y, [Y, \cdots Y]]}_{s_n} \cdots].$$

Remark 5.2 Later, we will use the following special version of BCH formula:

$$e^X Y e^{-X} = e^{[X, -]}(Y) := Y + [X, Y] + \frac{1}{2}[X, [X, Y]] + \cdots.$$

5.3 DT/PT Correspondence

The notion of stable pairs was introduced by Pandharipande–Thomas [146] in order to give a geometric understanding of the quotient series in Theorem 5.3. In [146], it was conjectured that each coefficient of the quotient series in Theorem 5.3 is an invariant counting stable pairs. Their conjecture was later proved in [31, 169] using a wall-crossing argument.

Let X be a smooth projective CY 3-fold. A Pandharipande–Thomas (PT) stable pair is defined as follows:

Definition 5.1 (*[146]*) A *stable pair* on X consists of a pair (F, s), where F is a pure one dimensional coherent sheaf and $s \colon \mathcal{O}_X \to F$ is surjective in dimension one, i.e. $\mathrm{Cok}(s)$ is at most zero-dimensional.

For $n \in \mathbb{Z}$ and $\beta \in H_2(X, \mathbb{Z})$, we have the moduli space of stable pairs

$$P_n(X, \beta) := \{\text{stable pairs } (F, s) \text{ with } [F] = \beta, \chi(F) = n\}.$$

Here $[F]$ is the homology class of the fundamental one cycle of F (see (7.3)). In [146], it is proved that $P_n(X, \beta)$ is a projective scheme such that the morphism of stacks

$$[P_n(X, \beta)/\mathbb{C}^*] \longrightarrow \widetilde{\mathcal{M}}_X(v), \ (F, s) \longmapsto (\mathcal{O}_X \xrightarrow{s} F)$$

is an open immersion. Here $\mathbb{C}^*$ acts on $P_n(X, \beta)$ trivially, the morphism of stabilizer subgroups $\mathbb{C}^* \to \mathrm{Aut}(\mathcal{O}_X \to F)$ is given by the scalar multiplication, v is the Chern character (1.22) and $\widetilde{\mathcal{M}}_X(v)$ is the moduli stack in Section 4.5. Therefore $P_n(X, \beta)$ admits a symmetric perfect obstruction theory and a zero-dimensional virtual fundamental class (see [146, Theorem 2.14]).

Definition 5.2 (*[146, Definition 2.16]*) The *Pandharipande–Thomas (PT) invariant* is defined by

$$P_{n,\beta} := \int_{[P_n(X,\beta)]^{\mathrm{vir}}} 1 \in \mathbb{Z}.$$

We form the generating series of PT invariants

$$P_\beta(X) := \sum_{n \in \mathbb{Z}} P_{n,\beta} q^n, \ P(X) := \sum_{\beta \geq 0} P_\beta(X) t^\beta.$$

Recall the generating series $I_\beta(X)$ of rank one DT invariants in (1.23). The *DT/PT correspondence* is stated as follows:

Theorem 5.2 ([31, 169]) *We have the identity of the Laurent series*

$$\frac{I_\beta(X)}{I_0(X)} = P_\beta(X).$$

Note that $I_n(X, \beta)$ parametrizes ideal sheaves, equivalently pairs (F, s) where F is a one-dimensional sheaf and $s \colon \mathcal{O}_X \to F$ is a surjection. The difference from stable pairs is summarized in the following table (Table 5.1).

Table 5.1 Difference of ideal sheaves and stable pairs

	Ideal sheaf	Stable pair
Section s	Surjective	Generically surjective
Target sheaf F	One dimensional	Pure one dimensional

Below we give an outline of how to derive the formula in Theorem 5.2 using the wall-crossing argument in the previous section. Let

$$\mathcal{Q}_n(X, \beta) \subset \widetilde{\mathcal{M}}_X(v)$$

be the substack of two term complexes $(\mathcal{O}_X \xrightarrow{s} F)$ such that F is one dimensional and s is surjective in dimension one. Here v is the Chern character (1.22). We have the inclusions

$$[P_n(X, \beta)/\mathbb{C}^*] \subset \mathcal{Q}_n(X, \beta) \supset [I_n(X, \beta)/\mathbb{C}^*].$$

For a moment, suppose that there exists $\sigma = (Z, \mathcal{A}) \in \text{Stab}^\circ(X)$ such that $\mathcal{M}_X^{\sigma\text{-ss}}(v) = \mathcal{Q}_n(X, \beta)$. For a geometric point of $\mathcal{Q}_n(X, \beta)$ corresponding to $(\mathcal{O}_X \xrightarrow{s} F)$, we have the distinguished triangle

$$(\mathcal{O}_X \twoheadrightarrow F') \longrightarrow (\mathcal{O}_X \longrightarrow F) \longrightarrow T[-1]. \qquad (5.8)$$

Here F' is the image of s and T is the cokernel of s which is a zero dimensional sheaf. Similarly we have the distinguished triangle

$$T'[-1] \longrightarrow (\mathcal{O}_X \longrightarrow F) \longrightarrow (\mathcal{O}_X \longrightarrow F''). \qquad (5.9)$$

Here $T' \subset F$ is the maximal zero dimensional subsheaf and $F'' = F/T'$. Note that $(\mathcal{O}_X \to F'')$ is a stable pair.

Now suppose that the change of DT/PT moduli spaces is given by wall-crossing at σ, i.e. there exist perturbations $\sigma^\pm$ of σ such that

$$\mathcal{M}_X^{\sigma^+\text{-ss}}(v) = \mathcal{M}_X^{\sigma^+\text{-st}}(v) = [I_n(X, \beta)/\mathbb{C}^*], \qquad (5.10)$$
$$\mathcal{M}_X^{\sigma^-\text{-ss}}(v) = \mathcal{M}_X^{\sigma^-\text{-st}}(v) = [P_n(X, \beta)/\mathbb{C}^*],$$
$$\mathcal{M}_X^{\sigma\text{-ss}}(0, 0, 0, -n) = \mathcal{M}_X^{\sigma^\pm\text{-ss}}(0, 0, 0, -n) = \mathcal{M}_n$$

and the sequences (5.8), (5.9) are Harder–Narasimhan filtrations in σ^+-stability, σ^--stability, respectively. Here $\mathcal{M}_n$ is the moduli stack of objects $T[-1]$ for a zero-dimensional sheaf T of length n. Similarly to the previous section, we set the following formal sum of δ-functions:

$$\delta_\beta^{\sigma^\pm\text{-ss}} := \sum_{n\in\mathbb{Z}} \delta^{\sigma^\pm\text{-ss}}(1, 0, -\beta, -n), \quad \delta_0 := \sum_{n\in\mathbb{Z}_{\geq 0}} \delta^{\sigma\text{-ss}}(0, 0, 0, -n),$$

which make sense in a suitable completion of the motivic Hall algebra of $\mathscr{A}$. Then similarly to (5.3), the sequences (5.8), (5.9) yield the identities

$$\delta_\beta^{\sigma\text{-ss}} = \delta_\beta^{\sigma^+\text{-ss}} * \delta_0 = \delta_0 * \delta_\beta^{\sigma^-\text{-ss}}.$$

Then by setting $\epsilon_0 = \log \delta_0$, we obtain the identity

$$\delta_\beta^{\sigma^+\text{-ss}} = \exp(\epsilon_0) * \delta_\beta^{\sigma^-\text{-ss}} * \exp(-\epsilon_0).$$

Using the formula in Remark 5.2, one can write the above identity in terms of Lie brackets:

$$\delta_\beta^{\sigma^+\text{-ss}} = \exp([\epsilon_0, -])(\delta_\beta^{\sigma^-\text{-ss}}),$$
$$= \delta_\beta^{\sigma^-\text{-ss}} + [\epsilon_0, \delta_\beta^{\sigma^-\text{-ss}}] + \frac{1}{2}[\epsilon_0, [\epsilon_0, \delta_\beta^{\sigma^-\text{-ss}}]] + \cdots.$$

Then by multiplying $(\mathbb{L} - 1)$, applying the integration map in Theorem 2.1 and noting (5.10), an easy calculation shows that (see the proof of [185, Theorem 3.17])

$$I_{n,\beta} = \sum_{k\geq 0} \frac{1}{k!} \sum_{n_0+n_1+\cdots+n_k=n} \left(\prod_{i=1}^k (-1)^{n_i-1} n_i N_{n_i}\right) P_{n_0,\beta}.$$

Here $N_n \in \mathbb{Q}$ is the generalized DT invariant (2.16) counting length n zero dimensional sheaves on X. In terms of generating series, we have

$$\sum_{n\in\mathbb{Z}} I_{n,\beta} q^n = \exp\left(\sum_{n\geq 1}(-1)^{n-1} n N_n q^n\right)\left(\sum_{n\in\mathbb{Z}} P_{n,\beta} q^n\right). \tag{5.11}$$

By substituting the formula for N_n in (2.16), we obtain the desired identity in Theorem 5.2.

Remark 5.3 In the above arguments, we assumed the existence of stability conditions which realize DT/PT wall-crossing. For a general CY 3-fold it is not known whether such stability conditions exist or not. Instead in [169], we introduced the notion of 'weak' stability conditions on $D^b(X)$, which are easier to construct than Bridgeland stability conditions, but enough to realize the above arguments. Also see [31, 185] for alternative arguments.

5.4   Rationality of PT Generating Series

In the rest of this chapter, we will give several other applications of wall-crossing formulas to generating series of DT invariants on CY 3-folds.

Using a similar argument of the proof of DT/PT correspondence, we can also prove the following theorem which is relevant for the rationality of the generating series.

Theorem 5.3 ([31, 170]) *For each $\beta \in H_2(X, \mathbb{Z})$ and $n \in \mathbb{Z}$, there exist invariants $N_{n,\beta} \in \mathbb{Q}$, $L_{n,\beta} \in \mathbb{Z}$ satisfying:*

(i) $N_{n,\beta} = N_{-n,\beta}$, $L_{n,\beta} = L_{-n,\beta}$,

(ii) $N_{n,\beta} = N_{n+\beta \cdot H, \beta}$ for any divisor H on X,

(iii) $L_{n,\beta} = 0$ for $|n| \gg 0$

such that we have the following formula:

$$P(X) = \prod_{n \geq 1} \exp((-1)^{n-1} N_{n,\beta} q^n t^\beta)^n \cdot \left(\sum_{n,\beta} L_{n,\beta} q^n t^\beta \right). \qquad (5.12)$$

The invariants $N_{n,\beta}$, $L_{n,\beta}$ are given as follows:

- The invariant $N_{n,\beta} \in \mathbb{Q}$ is the generalized DT invariant of one-dimensional semistable sheaves F on X whose Chern character is $(0, 0, \beta, n)$, i.e. $N_{n,\beta} = \mathrm{DT}_H(0, 0, \beta, n)$, which is shown to be independent of H (see [92, Theorem 6.16]).
- The invariant $L_{n,\beta} \in \mathbb{Z}$ is a DT invariant counting certain semistable objects in $D^b(X)$, which is self-dual with respect to the dualizing functor $\mathbf{R} \mathscr{H}om(-, \mathscr{O}_X)$.

The symmetry in (i) is a consequence of the self-duality of relevant stability conditions. The formula (5.12) is obtained by iterative applications of wall-crossing formulas, from a (weak) stability condition where PT stable pairs are stable objects, to a (weak) stability condition which is self-dual and giving the invariant $L_{n,\beta}$. An outline of the proof is also available in [172, Sect. 5].

The rationality of the PT generating series, conjectured in [146, Conjecture 3.2], is an easy consequence of the above formula.

Corollary 5.1 *The generating series $P_\beta(X)$ is the Laurent expansion of a rational function of q, invariant under $q \mapsto 1/q$.*

Remark 5.4 The result of Theorem 5.3 is a consequence of Theorem 5.2 and Corollary 5.1.

In [146, 172], a stronger rationality property of the series $P(X)$ was proposed:

Conjecture 5.1 ([146, Conjecture 3.14], [172, Conjecture 6.2]) There exist integers $n_{g,\beta} \in \mathbb{Z}$ for $g \geq 0$ and $\beta \in H_2(X, \mathbb{Z})$ such that we have

$$P(X) = \prod_{\beta>0} \left(\prod_{j=1}^{\infty} (1 - (-q)^j t^{\beta})^{jn_{0,\beta}} \cdot \prod_{g=1}^{\infty} \prod_{k=0}^{2g-2} (1 - (-q)^{g-1-k} t^{\beta})^{(-1)^{k+g} n_{g,\beta} \binom{2g-2}{k}} \right).$$

Remark 5.5 The integers $n_{g,\beta}$ should be Gopakumar–Vafa invariants discussed in Chap. 7. See Conjecture 7.2.

The result of Theorem 5.3 reduces Conjecture 5.1 to a conjectural property of the invariants $N_{n,\beta} \in \mathbb{Q}$.

Conjecture 5.2 ([92, Conjecture 6.20], [172, Conjecture 6.3]) We have the identity

$$N_{n,\beta} = \sum_{k \geq 1, k|(\beta,n)} \frac{1}{k^2} N_{1,\beta/k}.$$

In other words, the BPS invariant $\Omega_H(0, 0, \beta, n)$ in Conjecture 2.1 is independent of n.

Indeed, Theorem 5.3 implies that Conjectures 5.1 and 5.2 are equivalent (see [172, Theorem 6.4]). Conjecture 5.2 is still open in many cases.

5.5 Flop Formula of DT Invariants

Here we discuss the flop transformation formula of DT invariants. Let X be a smooth projective CY 3-fold. A projective birational contraction $f: X \to Y$ is a *flopping contraction* if f is an isomorphism in codimension one and the relative Picard number of f is one. A diagram

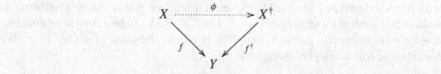

is called a *flop* of f if $f^{\dagger}$ is also a flopping contraction, and the birational map ϕ is not an isomorphism. It is a well-known fact from birational geometry [106] that any birational map $X \dashrightarrow X'$ between CY 3-folds is decomposed into a finite sequence of flops.

We define the following sub-generating series of $I(X)$:

$$I(X/Y) := \sum_{f_*\beta=0} I_{n,\beta}q^n t^\beta.$$

The following result gives a flop transformation formula of curve-counting DT invariants.

Theorem 5.4 ([38, 175]) *We have the identities of the generating series*

$$I(X^\dagger/Y) = i \circ \phi_* I(X/Y), \quad \frac{I(X^\dagger)}{I(X^\dagger/Y)} = \phi_* \frac{I(X)}{I(X/Y)}.$$

Here ϕ_*, i are variable changes $\phi_*(\beta, n) = (\phi_*\beta, n)$, $i(\beta, n) = (-\beta, n)$, where $\phi_*\beta$ is determined by $\phi_*\beta \cdot D = \beta \cdot \phi_*^{-1}D$ for any divisor D on $X^\dagger$ and $\phi_*^{-1}D$ is the strict transform.

In particular, if the exceptional loci of f, $f^\dagger$ are $(-1, -1)$-curves C, $C^\dagger$, from the formula (1.25) we have

$$I(X^\dagger) = \phi_* I(X) \cdot \prod_{n \geq 1}(1 - (-q)^n t^{[C^\dagger]}) \cdot \prod_{n \geq 1}(1 - (-q)^n t^{-[C^\dagger]}).$$

The result of Theorem 5.4 is proved using derived equivalence $D^b(X^\dagger) \overset{\sim}{\to} D^b(X)$ proved in [26, 191], and the strategy described in Sect. 5.1. In the case of a flop of the $(-1, -1)$-curve, there is an alternative approach using the degeneration formula [78].

5.6 Jacobi Property for Elliptic CY 3-Folds

Let X be a smooth projective CY 3-fold which admits an elliptic fibration

$$\pi: X \longrightarrow S$$

with fibers reduced and irreducible curves of arithmetic genus one. We furthermore assume that π admits a section $\iota: S \to X$. Let $C \in H_2(S, \mathbb{Z})$ be a class of an effective divisor with arithmetic genus h, and $F \in H_2(X, \mathbb{Z})$ the class of a fiber of π. We consider the following generating series:

$$P_C(q, t) := \sum_{d \geq 0, n \in \mathbb{Z}} P_{n, \iota_*C + dF} q^n t^d.$$

When $C = 0$, as an application of the formula (5.12), we have the complete evaluation of $P_0(q, t)$.

Theorem 5.5 ([172, Theorem 6.9]) *We have the identity*

$$P_0(q, t) = \prod_{l,m \geq 1} (1 - (-q)^l t^m)^{-l \cdot e(X)} \cdot \prod_{m \geq 1} (1 - t^m)^{-e(S)}.$$

Let $Z_C(q, t)$ be the quotient series

$$Z_C(q, t) = \frac{P_C(q, t)}{P_0(q, t)}.$$

Suppose that C is reduced, i.e. every effective divisor on S with class C is a reduced divisor. In this case, the following result is proved by Oberdieck–Shen [138].

Theorem 5.6 *([138, Theorem 1]) Let C be reduced and of arithmetic genus h. Then we have the following functional equation of the generating series:*

$$Z_C(q^{-1}t, t) = q^{2(h-1)} t^{-(h-1)} Z_C(q, t).$$

The result of Theorem 5.6 is also proved along with the idea in Section 4.6, where the equivalence Φ is an autoequivalence given by relative Fourier–Mukai transforms.

By Corollary 5.1, the identity $Z_C(q^{-1}, t) = Z_C(q, t)$ holds as an element in $\mathbb{Q}(q)[\![t]\!]$. Here we note that the above identity does not hold as a generating series, and only hold as rational functions in q-variables. By combining with Theorem 5.6, we obtain the identity in $\mathbb{Q}(q)[\![t]\!]$ for $\lambda \in \mathbb{Z}$ (see [138, Corollary 1])

$$Z_C(qt^{\lambda}, t) = q^{-2(h-1)\lambda} t^{-(h-1)\lambda^2} Z_C(q, t).$$

By the arguments from physics, it is conjectured by Huang–Katz–Klemm [80] that the series $Z_C(q, t)$ is a meromorphic Jacobi form of index $h - 1$. The above formula gives an elliptic transformation law of the conjectured Jacobi form property for $Z_C(q, t)$.

5.7 PT Invariants for Local K3 Surfaces

Let S be a smooth projective K3 surface, i.e. dim $S = 2$, $K_S \cong \mathcal{O}_S$ and $H^1(\mathcal{O}_S) = 0$, e.g. a smooth quartic surface $S \subset \mathbb{P}^3$. Let X be the total space of its canonical line bundle

$$X = \mathrm{Tot}_S(K_S) = S \times \mathbb{A}^1.$$

Let us take $\beta \in H_2(S, \mathbb{Z}) = H_2(X, \mathbb{Z})$ and $n \in \mathbb{Z}$. The moduli space of stable pairs $P_n(X, \beta)$ is not compact, but one can define the PT invariant by taking the integration of the Behrend function

$$P_{n,\beta} := \int_{P_n(X,\beta)} \chi_B \, de.$$

The generating series of the above PT invariants is completely known by combining the results of several references:

Theorem 5.7 ([127, 148, 173]) *The generating series $P(X)$ is given by*

$$P(X) = \prod_{r\geq 0, \beta>0, n\geq 0} (1 - (-q)^n t^\beta)^{-(n+2r)c(\beta^2/2-r(n+r)+1)}$$

$$\cdot \prod_{r>0, \beta>0, n>0} (1 - (-q)^{-n} t^\beta)^{-(n+2r)c(\beta^2/2-r(n+r)+1)}.$$

Here $c(m) \in \mathbb{Z}$ is determined by the identity

$$\sum_{m\in\mathbb{Z}} c(m) q^m = \prod_{m\geq 1} (1 - q^m)^{-24}.$$

The formula in Theorem 5.7 is a combination of the wall-crossing formula proved in [173, Theorem 5.5]

$$P(X) = \prod_{r\geq 0, \beta>0, n\geq 0} \exp(J(r, \beta, r+n)(-q)^n t^\beta)^{-(n+2r)}$$

$$\cdot \prod_{r>0, \beta>0, n>0} \exp(J(r, \beta, r+n)(-q)^{-n} t^\beta)^{-(n+2r)} \qquad (5.13)$$

with the calculation of $J(r, \beta, n) \in \mathbb{Q}$ in [127, Corollary 6.10], using the solution of the Katz–Klemm–Vafa formula [96] by Pandharipande–Thomas [148, Theorem 1].

Remark 5.6 In [173, Theorem 5.5], the formula (5.13) is proved for naive Euler characteristics of stable pair moduli spaces. The Behrend-weighted version is now available by involving [185, Theorem 1.1] in the argument of [173].

Remark 5.7 For when β is an irreducible curve class, the naive Euler characteristics of stable pair moduli spaces on S was calculated by Kawai–Yoshioka [97, Theorem 5.80]. The proof of the formula 5.13 is obtained by interpreting the arguments in [97, Sect. 5] in terms of wall-crossing in the derived category of X.

In the formula (5.13), the invariant $J(v) \in \mathbb{Q}$ is the generalized DT invariant on X, counting compactly supported semistable sheaves E on X such that $\pi_* E$ has Mukai vector v, where $\pi : X \to S$ is the projection. Namely for $v = (r, \beta, n)$, the above condition is

$$\mathrm{ch}(\pi_* E)\sqrt{\mathrm{td}_S} = (r, \beta, n) \in H^0(S) \oplus H^2(S) \oplus H^4(S).$$

If $v \in H^{2*}(S, \mathbb{Z})$ is primitive, then the moduli space which defines $J(v)$ is $M_S(v) \times \mathbb{A}^1$, where $M_S(v)$ is the moduli space of stable sheaves on S with Mukai vector v. It is well-known that $M_S(v)$ is a holomorphic symplectic manifold which is deformation equivalent to the Hilbert scheme of points on S with dimension $(v, v) + 2$, where $(-, -)$ is the Mukai pairing on $H^{2*}(S, \mathbb{Z})$. So in this case, we have

$$J(v) = -e(M_S(v)) = -e(\text{Hilb}^{(v,v)/2+1}(S)). \tag{5.14}$$

The Euler numbers of Hilbert schemes of points are determined by the Göttsche formula [72]

$$\sum_{n\geq 0} e(\text{Hilb}^n(S))q^n = \prod_{n\geq 1}(1 - q^n)^{-24}. \tag{5.15}$$

If v is not necessarily primitive, it is proved in [127, Corollary 6.10] that the following formula holds:

$$J(v) = - \sum_{k\geq 1, k|v} \frac{1}{k^2} e(\text{Hilb}^{(v/k,v/k)/2+1}(S)). \tag{5.16}$$

The naive Euler characteristic version of the above formula was conjectured in [173, Conjecture 1.3], and is also proved in [127, Corollary 6.10]. If $r = 0$, the above formula is nothing but the multiple cover formula in (5.2) by (5.14). The formula (5.2) for the local K3 surface is proved in [127, Corollary 6.8] by showing the Gopakumar–Vafa form of the series $P(X)$ in Conjecture 5.1, which is implied by the comparison with reduced PT invariants together with the study of the latter invariants in [148]. The formula (5.16) for an arbitrary v is reduced to the $r = 0$ case in [173, Section 6] by deformation argument together with Fourier–Mukai transforms.

The formula in Theorem 5.7 is obtained by substituting (5.16), (5.15) into the formula (5.13).

5.8 Other Applications

There exist several other references on applications of wall-crossing formulas to generating series of DT invariants. Here we list some of them.

- **(Vafa–Witten S-duality conjecture)**
  The generating series of DT invariants counting two-dimensional torsion sheaves on a CY 3-fold is expected to be a Jacobi form. This may be regarded as a CY 3-fold version of the Vafa–Witten S-duality conjecture for surfaces [190] (see [160, 161] for recent mathematical foundation of Vafa–Witten theory). The flop transformation formula for generating series of such invariants is obtained in [177] via wall-crossing, and proves that the error terms are described by a Jacobi form. The

wall-crossing formula is also used in [124, 180] to determine the above series for local $\mathbb{P}^2$.

- **(Ooguri–Strominger–Vafa conjecture)**
  In physics, Denef–Moore [53] used a wall-crossing argument to derive the Ooguri–Strominger–Vafa (OSV) conjecture relating black hole entropy and topological string. In mathematics, this is a relationship between the generating series of DT invariants for two-dimensional torsion sheaves on CY 3-folds and rank one DT invariants counting curves or stable pairs. In [174], it is proved that Denef–Moore's argument is mathematically justified if we assume the BG-type inequality conjecture in Conjecture 4.2. If the surface class inside CY 3-fold is sufficiently large, the above relationship can be much simplified as shown in [60].

- **(Crepant resolution conjecture)**
  For a finite subgroup $G \subset SL_3(\mathbb{C})$, the quotient stack $\mathscr{Y} = [\mathbb{C}^3/G]$ is a smooth Deligne–Mumford stack with a singular coarse moduli space $\mathscr{Y} \to Y$. The derived McKay correspondence in [32] proves the existence of a crepant resolution $X \to Y$ together with a derived equivalence between X and $\mathscr{Y}$. The *crepant resolution conjecture* by Bryan–Cadman–Young [33] is a conjectural relation between curve-counting DT invariants on X and those on $\mathscr{Y}$ when G satisfies the Hard–Lefschetz condition (which is equivalent to saying that the fibers of $X \to Y$ have at most one dimension). By wall-crossing, the former is related to *Bryan–Steinberg (BS) pair invariants* [36], which are closely related to the DT invariants on $\mathscr{Y}$ under derived equivalence. The original crepant resolution conjecture is now proved by further wall-crossing of BS pairs by Beentjes–Calabrese–Rennemo [9] (with a modification of the statement as an identity of rational functions rather than generating series). Where G does not satisfy the Hard–Lefschetz condition, a simple relation between DT invariants on X and $\mathscr{Y}$ has not yet been found. In the case of $G = \mu_3$, an intricate relationship is described in [179] via wall-crossing.

- **(DT invariants on abelian 3-folds)**
  Let A be an abelian 3-fold. The usual DT invariants on A are not interesting since they are zero. Instead one can define the *reduced DT invariants* on abelian 3-folds, defined by Gulbrandsen [74] by taking the quotient of the relevant moduli spaces by the action of $A \times \widehat{A}$, where $\widehat{A}$ is the dual abelian 3-fold. In [137], the wall-crossing formula of reduced DT invariants on abelian 3-folds is investigated with respect to Bridgeland stability conditions on them. As a result, there is a strong constraint of reduced DT invariants induced by the action of the autoequivalence group of $D^b(A)$. On the other hand, Bryan–Oberdieck–Pandharipande–Yin [35] proposed the conjecture describing curve-counting reduced DT invariants on abelian 3-folds. The constraints obtained in [137] are compatible with their conjecture.

- **(Higher rank DT invariants)**
  As we observed in Sect. 1.5, the rank one DT invariants count curves or points inside CY 3-folds. We can also consider the DT invariants counting higher rank stable sheaves. If the rank and degree are coprime, their generating series satisfy properties similar to rank one DT invariants, e.g. DT/PT correspondence, rationality [67, 185]. On the other hand, if rank and degree are not coprime, their structure seems to be more complicated. See [157, 171] for rank two and three versions

of the formula (1.24). Recently Feyzbakhsh–Thomas have shown the existence of universal formulae describing rank r DT invariants for $r > 0$ in terms of rank zero DT invariants [61], and rank one DT invariants [62], assuming the BG-type inequality conjecture in Conjecture 4.2. Their method is to use rank reduction using JS stable pairs and wall-crossing.

Chapter 6
Cohomological Donaldson-Thomas Invariants

Abstract The DT invariants counting stable sheaves are integers and coincide with Behrend-weighted Euler characteristics of the moduli spaces. As these are closely related to topological Euler numbers of the moduli spaces, one may ask a question whether the DT invariants can be refined to some cohomology theory whose Euler characteristics recover the DT invariants. In this chapter, we explain how to establish such a theory via Joyce's d-critical structures.

6.1 Vanishing Cycle Sheaves

Let M be a moduli space of stable sheaves (or stable objects in the derived category) on a smooth projective CY 3-fold X. Recall that the associated DT invariant is defined either in the following ways:

  (i) integral of the zero dimensional virtual fundamental cycle on M,
  (ii) integral of the Behrend constructible function on M.

The latter construction provides a hint to refine DT invariants to a cohomological level, i.e. some cohomology theory of M whose Euler characteristic recovers the DT invariant. Indeed, if M is smooth, the DT invariant is the signed Euler characteristic $(-1)^{\dim M} e(M)$, and the shifted cohomology $H^{*+\dim M}(M, \mathbb{Q})$ recovers the DT invariant by taking its Euler characteristic.

   An idea of such a refinement is to interpret Behrend function in terms of pointwise Euler characteristics of some perverse sheaf on M. Here we recall the definition of perverse sheaves (see [17] for details on the theory of perverse sheaves):

Definition 6.1 Let M be a $\mathbb{C}$-scheme and $D^b_{\mathrm{con}}(M)$ the derived category of constructible sheaves on M with $\mathbb{Q}$-coefficients. An object $\mathscr{E}$ in $D^b_{\mathrm{con}}(M)$ is called a *perverse sheaf* (w.r.t. the middle perversity) if the following conditions are satisfied for all $i \in \mathbb{Z}$:

$$\dim \mathrm{Supp}\,\mathscr{H}^i(\mathscr{E}) \leq -i, \ \dim \mathrm{Supp}\,\mathscr{H}^i(\mathbb{D}_M(\mathscr{E})) \leq -i.$$

© The Author(s), under exclusive license to Springer Nature Singapore Pte Ltd. 2021   71
Y. Toda, *Recent Progress on the Donaldson—Thomas Theory*,
SpringerBriefs in Mathematical Physics 43,
https://doi.org/10.1007/978-981-16-7838-7_6

Here $\mathbb{D}_M$ is the Verdier duality functor.

The subcategory of perverse sheaves on M is denoted by

$$\mathrm{Perv}(M) \subset D^b_{\mathrm{con}}(M).$$

It is the heart of a t-structure on $D^b_{\mathrm{coh}}(M)$, in particular, it is an abelian subcategory.

Example

For an irreducible variety M, we have the *intersection complex*

$$\mathrm{IC}(M) = j_{!*}\mathbb{Q}_U[n] \in \mathrm{Perv}(M).$$

Here $j : U \hookrightarrow M$ is the smooth locus and $n = \dim M$. The intersection complex is introduced in [71], and coincides with the above object by [17, Proposition 2.1.11]. In particular, if M is smooth, we have $\mathrm{IC}(M) = \mathbb{Q}_M[n]$.

Let V be a smooth $\mathbb{C}$-scheme and $f : V \to \mathbb{C}$ be a morphism such that the critical locus $\mathrm{Crit}(f) \subset V$ is set theoretically supported on $V_0 := f^{-1}(0)$. Below we explain the vanishing cycle sheaf associated with (V, f), whose basic reference is [56, Sect. 4.2].

Let $V^0 := f^{-1}(\mathbb{C}^*)$ and $\mathbb{C} \to \mathbb{C}^*$ the universal cover. We consider the following diagram:

$$\begin{array}{ccccccc}
V_0 & \overset{i}{\hookrightarrow} & V & \overset{j_1}{\longleftarrow} & V^0 & \overset{j_2}{\longleftarrow} & \widetilde{V}^0 \\
\downarrow & & \downarrow f & & \downarrow f|_{V^0} & & \downarrow \\
\{0\} & \hookrightarrow & \mathbb{C} & \longleftarrow & \mathbb{C}^* & \longleftarrow & \mathbb{C}
\end{array}$$

Here every square is Cartesian. We set $j = j_1 \circ j_2 : \widetilde{V}^0 \to V$. Then for each $F \in D^b_{\mathrm{con}}(V)$ we have the following object:

$$\Psi_f(F) := i^{-1}\mathbf{R}j_* j^{-1}(F) \in D^b_{\mathrm{con}}(V_0).$$

By the adjunction of $\mathbf{R}j_*$ and j^{-1}, there exists an object $\Phi_f(F) \in D^b_{\mathrm{con}}(V_0)$ and the following distinguished triangle:

$$i^{-1}F \longrightarrow \Psi_f(F) \longrightarrow \Phi_f(F). \tag{6.1}$$

We set $\Psi^p_f(-) := \Psi_f(-)[-1]$ and $\Phi^p_f(-) := \Phi_f(-)[-1]$. Then they determine the functors (see [56, Theorem 5.2.21])

$$\Psi_f^p, \Phi_f^p : \mathrm{Perv}(V) \longrightarrow \mathrm{Perv}(V_0)$$

called the *nearby cycle functor* and *vanishing cycle functor*, respectively. The object

$$\phi_f := \Phi_f^p(\mathbb{Q}_V[\dim V]) \in \mathrm{Perv}(V_0)$$

can be shown to be supported on $\mathrm{Crit}(f) \subset V_0$, so is regarded as a perverse sheaf on $\mathrm{Crit}(f)$. The object ϕ_f is called a *vanishing cycle sheaf*.

Example

Let $V = \mathbb{C}$ and $f(z) = z^{n+1}$. Then $\mathrm{Crit}(f) = \{z^n = 0\} \subset \mathbb{C}$, and the vanishing cycle sheaf is $\phi_f = \mathbb{Q}_0^{\oplus n}$.

For each $x \in \mathrm{Crit}(f)$, from the definitions of Milnor fibers (1.19) and vanishing cycle sheaves (6.1) we have

$$\chi(\phi_f|_x) = (-1)^{\dim V}(1 - e(M_f(x))),$$

i.e. $\chi(\phi_f|_-)$ is nothing but the Behrend function χ_B on $\mathrm{Crit}(f)$. By taking a stratification of $\mathrm{Crit}(f)$ where the Behrend function is constant on each strata, the above fact easily shows that (see [24, Remark 6.14])

$$\chi(H^*(\mathrm{Crit}(f), \phi_f)) = \int_{\mathrm{Crit}(f)} \chi_B \, de,$$

i.e. the cohomology theory $H^*(\mathrm{Crit}(f), \phi_f)$ refines the integration of the Behrend function.

6.2 D-critical Loci

The moduli space M of stable sheaves on a CY 3-fold is locally a critical locus. So we can associate the perverse sheaf of vanishing cycles, locally at each point of M. However, it is not obvious whether these locally defined perverse sheaves glue to give a global perverse sheaf. At least, we need to know how local critical charts (V, f) are related on overlaps. A d-critical structure introduced by Joyce [90] provides such information.

For any $\mathbb{C}$-scheme M, Joyce [90, Theorem 2.1 (i)] shows that there exists a canonical sheaf of $\mathbb{C}$-vector spaces $\mathcal{S}_M$ on M satisfying the following: for any Zariski open subset $U \subset M$ and a closed embedding $U \subset V$ for a smooth $\mathbb{C}$-scheme V, there is an exact sequence

$$0 \longrightarrow \mathcal{S}_M|_U \longrightarrow \mathcal{O}_V/I^2 \xrightarrow{d_{\mathrm{DR}}} \Omega_V/I\Omega_V.$$

Here $I \subset \mathcal{O}_V$ is the ideal sheaf which defines U and d_{DR} is the de-Rham differential. Moreover, there is a natural decomposition $\mathcal{S}_M = \mathcal{S}_M^0 \oplus \mathbb{C}_M$, where $\mathbb{C}_M$ is the constant sheaf on M (see [90, Theorem 2.1 (ii)]). The sheaf $\mathcal{S}_M^0$ restricted to U is the kernel of the composition

$$\mathcal{S}_M|_U \hookrightarrow \mathcal{O}_V/I^2 \longrightarrow \mathcal{O}_{U^{\mathrm{red}}}.$$

Example

Suppose that $f: Y \to \mathbb{C}$ is a function such that

$$U = \mathrm{Crit}(f), \quad f|_{U^{\mathrm{red}}} = 0. \tag{6.2}$$

Then $I = (df)$ and $f + (df)^2$ is an element of $\Gamma(U, \mathcal{S}_M^0|_U)$.

A d-critical locus is defined as follows:

Definition 6.2 ([90, Definition 2.5]) A pair (M, s) for a $\mathbb{C}$-scheme M and $s \in \Gamma(M, \mathcal{S}_M^0)$ is called a *d-critical locus* if for any $x \in M$, there exists a Zariski open neighborhood $x \in U \subset M$, a closed embedding $i: U \hookrightarrow Y$ into a smooth scheme Y, an element $f \in \Gamma(\mathcal{O}_Y)$ satisfying (6.2) such that $s|_U = f + (df)^2$ holds. In this case, the data

$$(U, Y, f, i) \tag{6.3}$$

is called a *d-critical chart*. The section s is called a *d-critical structure* of M.

Remark 6.1 If M is smooth, then $\mathcal{S}_M^0 = 0$ so there is a unique (trivial) choice of its d-critical structure, $s = 0$.

6.3 Gluing of Vanishing Cycle Sheaves

In the situation of Definition 6.2, let $\pi: E \to Y$ be a vector bundle with an element $q \in \Gamma(Y, \mathrm{Sym}^2(E^\vee))$ giving non-degenerate quadratic forms on each fiber. Then we have a function on E

$$\widetilde{f} := \pi^* f + q: E \longrightarrow \mathbb{C}.$$

It is easy to see that the zero section $Y \to E$ restricts to the isomorphism $g: \mathrm{Crit}(f) \xrightarrow{\cong} \mathrm{Crit}(\widetilde{f})$, and the d-critical structures on $f + (df)^2$ and $\widetilde{f} + (d\widetilde{f})^2$

correspond to each other by the above isomorphism, i.e. $U = \mathrm{Crit}(f)$ admits two d-critical charts

$$(U, Y, f, \mathrm{id}), \; (U, E, \widetilde{f}, g). \tag{6.4}$$

Conversely, it is proven in [90] that any two local d-critical charts on M are 'essentially' related as in (6.4) (see [90, Sect. 2.3] for precise statements).

Analytic locally on Y, $\widetilde{f}$ is written as

$$\widetilde{f} : Y \times \mathbb{C}^r \longrightarrow \mathbb{C}, \; \widetilde{f}(y, z_1, \ldots, z_r) = f(y) + \sum_{i=1}^{r} z_i^2.$$

By the *Thom–Sebastiani isomorphism* (see [125, Theorem]), the vanishing cycle sheaf is described as

$$\phi_{\widetilde{f}} \cong \phi_f \boxtimes \phi_{z_1^2} \boxtimes \cdots \boxtimes \phi_{z_r^2}.$$

The vanishing cycle sheaf ϕ_{z^2} for the function $z^2 \colon \mathbb{C} \to \mathbb{C}$ is isomorphic to the constant sheaf supported on the origin. Therefore $\phi_{\widetilde{f}}$ and ϕ_f are isomorphic analytic locally on Y, non-canonically up to choices of trivializations $\phi_{z_i^2} \cong \mathbb{Q}_0$. Globally, this means that $\phi_{\widetilde{f}}$ and ϕ_f differ by a tensor product with a rank one local system on U.

Therefore for a d-critical locus (M, s), locally defined perverse sheaves of vanishing cycles just differ by a rank one local system on overlaps of an open cover. An idea of gluing these vanishing cycle sheaves is to modify ϕ_f with $\phi_f \otimes \mathscr{L}_U$ for some rank one local system $\mathscr{L}_U$ on U so that it cancels the difference of vanishing cycle sheaves on overlaps.

Let us consider two d-critical charts (6.4) again. The quadratic form q on E induces the isomorphism $q \colon E \xrightarrow{\cong} E^\vee$, which induces the isomorphism $\det q \colon \det(E)^{\otimes 2} \xrightarrow{\cong} \mathcal{O}_Y$. Since $N_{Y/E} = E$, the above isomorphism induces the isomorphism of squares of canonical line bundles,

$$K_E^{\otimes 2}|_U \xrightarrow{\cong} K_Y^{\otimes 2}|_U \otimes \det(N_{Y/E})|_U^{\otimes 2} \xrightarrow{\cong} K_Y^{\otimes 2}|_U.$$

The above observation leads to the following: given a d-critical locus (M, s), there exists a line bundle $K_{M,s}$ on M^{red} such that for any d-critical chart (6.3) we have a canonical isomorphism

$$K_{M,s}|_{U^{\mathrm{red}}} \xrightarrow{\cong} K_Y^{\otimes 2}|_{U^{\mathrm{red}}}. \tag{6.5}$$

Such a line bundle, and its existence is proved in [90, Theorem 2.28] is called a *virtual canonical line bundle*. The notion of orientation data is given as follows:

Definition 6.3 (*[90, Definition 2.31]*) For a d-critical locus (M, s), its *orientation data* is a line bundle $\sqrt{K_{M,s}} \in \mathrm{Pic}(M^{\mathrm{red}})$ together with an isomorphism

$$(\sqrt{K_{M,s}})^{\otimes 2} \xrightarrow{\cong} K_{M,s}. \tag{6.6}$$

Given the orientation data of (M, s), by composing (6.5) with (6.6) we have an isomorphism

$$(\sqrt{K_{M,s}})^{\otimes 2}|_{U^{\mathrm{red}}} \xrightarrow{\cong} K_Y^{\otimes 2}|_{U^{\mathrm{red}}}. \tag{6.7}$$

Let $\iota_U : \mathscr{P}_U \to U$ be the μ_2-torsor parameterizing local square roots of the isomorphism (6.7). Then we have a splitting $\iota_{U*} \mathbb{Q}_{\mathscr{P}_U} \cong \mathbb{Q}_U \oplus \mathscr{L}_U$ for a rank one local system $\mathscr{L}_U$ on U. It is proved in [24] that $\phi_f \otimes \mathscr{L}_U$ cancels out the difference of vanishing cycle sheaves, so that they glue to give a global perverse sheaf.

Theorem 6.1 ([24, Theorem 6.9]) *Given the orientation data of a d-critical locus* (M, s), *there exists a perverse sheaf* ϕ_M *on* M *such that for each d-critical chart* (U, Y, f, i) *there is a canonical isomorphism* $\phi_M|_U \xrightarrow{\cong} \phi_f \otimes \mathscr{L}_U$.

6.4   Cohomological DT Invariants

Let X be a smooth projective CY 3-fold and $M = M_X^{H\text{-st}}(v)$ is a fine moduli space of H-stable sheaves on X. Then we have the following:

Theorem 6.2 ([25, Corollary 6.7]) *There exists a canonical d-critical structure on* M.

Remark 6.2 From derived algebraic geometry, it is proved in [149, Theorem 0.1] that the moduli space M is the classical truncation of a derived moduli scheme with a (-1)-*shifted symplectic structure*. The result of [25, Theorem 6.6] states that the classical truncation of a (-1)-shifted symplectic derived scheme admits a canonical d-critical structure. As a result of Theorem 6.2, the moduli space M is Zariski locally written as a critical locus. Although the analytic local statement is proved in [92, Theorem 5.4] via gauge theory, the Zariski local result is proved using derived algebraic geometry.

For a fine moduli space of sheaves M on a CY 3-fold X, let s be its canonical d-critical structure in Theorem 6.2. Then for a universal sheaf $\mathcal{U} \in \mathrm{Coh}(X \times M)$, the virtual canonical line bundle is described as

$$K_{M,s} = K_M^{\mathrm{vir}} := \det \mathbf{R} p_{M*} \mathbf{R} \mathcal{H}om(\mathcal{U}, \mathcal{U}). \tag{6.8}$$

Here $p_M : X \times M \to M$ is the projection. This is because

$$(\det \mathrm{Ext}^1(E, E))^{\otimes 2} = \det \mathrm{Ext}^1(E, E) \otimes \det \mathrm{Ext}^2(E, E)^{-1}$$

by Serre duality $\mathrm{Ext}^2(E, E) \cong \mathrm{Ext}^1(E, E)^{\vee}$. It is proved in [136, Sect. 6.2] that K_M^{vir} has a square root, so there is orientation data on M. By Theorem 6.1, there is a global perverse sheaf ϕ_M on M. The cohomological DT invariant, which first appeared in Kontsevich–Soibelman's paper [109], is defined as follows (see [24, Remark 6.14]):

Definition 6.4 The *cohomological DT invariant* is the hypercohomology $\mathbb{H}^*(M, \phi_M)$ of the perverse sheaf ϕ_M.

The cohomological DT invariant recovers the numerical DT invariant by (see [24, Remark 6.14])

$$\mathrm{DT}_H(v) = \sum_{k \in \mathbb{Z}} (-1)^k \dim \mathbb{H}^k(M, \phi_M). \tag{6.9}$$

Remark 6.3 Contrary to the numerical DT invariant, the cohomological DT invariant depends on the choice of orientation data (see (6.10)). Recently, Joyce–Upmeier [93, Theorem 4.4] constructed canonical orientation data on M satisfying several expected properties. One can then use their canonical orientation data to define canonical cohomological DT invariants.

Example

Let $v = \mathrm{ch}(\mathcal{O}_x)$ for a point $x \in X$. Then there is a natural isomorphism

$$X \xrightarrow{\cong} M = M_H(v), \quad y \longmapsto \mathcal{O}_y.$$

As X is smooth, there is only a trivial d-critical structure $s = 0$ on M. The virtual canonical line bundle on M is identified with $K_X^{\otimes 2} \cong \mathcal{O}_X$, so we have an orientation data $K_X \cong \mathcal{O}_X$. The associated perverse sheaf ϕ_M is then $\mathbb{Q}_M[3]$, so the cohomological DT invariant in this case is $H^{*+3}(X, \mathbb{Q})$.

Example

Let $C \subset \mathbb{P}^2$ be defined by $C = \{zy^2 = x^3 + zx^2\}$ where $[x : y : z]$ is the homogeneous coordinate of $\mathbb{P}^2$. Note that C is a rational curve with one node at $p = [0 : 0 : 1]$. There is a scheme M with $M^{\mathrm{red}} = C$, smooth outside p and $\widehat{\mathcal{O}}_{M,p} = \mathbb{C}[\![u, v]\!]/(u^2 v, uv^2) = \mathrm{Crit}(u^2 v^2)$. There is a global critical description of M as follows. Let $\mathscr{C} \to T$ be a versal deformation of C, where $0 \in T \subset \mathbb{A}^1$ is a sufficiently small open subset and

$$\mathscr{C} = \{zy^2 = x^3 + zx^2 + tz^3\} \subset \mathbb{P}^2 \times T.$$

We define $f : \mathscr{C} \to \mathbb{C}$ to be $f([x : y : z], t) = t^2$. Then one can check that $M = \text{Crit}(f) \subset \mathscr{C}$. In particular, we have a d-critical structure $s = f + (df)^2$ on M.

The d-critical locus M should be obtained as a moduli of sheaves as follows. Let $C \subset X$ be an embedding into a CY 3-fold X such that the normal bundle is generic, as in the situation [34, Theorem 1.2]. Then we have a set theoretic bijection

$$C \longrightarrow M_X^{H\text{-st}}(0, 0, [C], 1), \quad x \longmapsto I_x^\vee,$$

where $I_x \subset \mathscr{O}_C$ is the ideal sheaf of x. However, the above map cannot be an isomorphism of schemes as C is not locally written as a critical locus (see Remark 1.12). Indeed, the moduli space $M_X^{H\text{-st}}(0, 0, [C], 1)$ should be isomorphic to M, though its rigorous proof is not available in references.

Since $K_\mathscr{C} = \mathscr{O}_\mathscr{C}$, we have $K_{M,s} = \mathscr{O}_C$. As $\text{Pic}^0(C) = \mathbb{C}^*$, we have two choices of orientation data: $\sqrt{K_{M,s}} = \mathscr{O}_C$ or L where L is a unique non-trivial 2-torsion element in $\text{Pic}^0(C)$. Let ϕ_M be the corresponding perverse sheaf in Theorem 6.1. It is an easy exercise to calculate the cohomological DT invariants as

$$\mathbb{H}^*(M, \phi_M) = \begin{cases} \mathbb{Q}[1] \oplus \mathbb{Q}^{\oplus 3} \oplus \mathbb{Q}[-1], & \text{if } \sqrt{K_{M,s}} = \mathscr{O}_C, \\ \mathbb{Q}, & \text{if } \sqrt{K_{M,s}} = L. \end{cases} \quad (6.10)$$

In particular, the cohomological DT invariant depends on the choice of orientation data.

Remark 6.4 The notion of a d-critical structure can be extended to an Artin stack $\mathscr{M}$ (see [90, Sect. 2.8]). Given orientation data on a d-critical stack $(\mathscr{M}, s)$, we also have the gluing of vanishing cycle sheaves $\phi_\mathscr{M}$ on $\mathscr{M}$ (see [18, Corollary 4.9]). If $\mathscr{M} = \mathscr{M}_X^{H\text{-ss}}(v)$ is the moduli stack of H-semistable sheaves, the direct sum of hypercohomologies $\mathbb{H}^*(\mathscr{M}, \phi_\mathscr{M})$ for $v \in \Gamma$ with a fixed reduced Hilbert polynomial is expected to carry an algebra structure, called *cohomological Hall algebra (COHA)* (see [93, Remark 4.15 (d)]). The COHA for a quiver with super-potential is constructed by Kontsevich–Soibelman [109] and Davison [51].

Remark 6.5 There is another type of refinement of DT theory, called *motivic DT invariant*. This is obtained by taking the motivic vanishing cycles [54] associated with d-critical charts (6.3), and glueing them using orientation data. The motivic DT theory first appeared in [108], and a rigorous foundation is obtained in [37]. Some explicit calculations are studied in [13, 133].

6.5 Dual Obstruction Cone

Let N be a scheme with a perfect obstruction theory

$$\mathscr{E} \longrightarrow \tau_{\geq -1} \mathbb{L}_N. \tag{6.11}$$

In this setting, there is a canonical way to construct a d-critical locus, called the dual obstruction cone [84].

Definition 6.5 ([84, Sect. 2]) The *dual obstruction cone M* of N is defined by

$$\pi : M := \mathrm{Spec}_N \mathrm{Sym}^{\bullet}(\mathscr{H}^1(\mathscr{E}^{\vee})) \longrightarrow N.$$

Remark 6.6 For any point $x \in N$, the fiber of $M \to N$ at x is identified with the vector space $\mathscr{H}^{-1}(\mathscr{E}|_x) = \mathscr{H}^1(\mathscr{E}^{\vee}|_x)^{\vee}$, the dual of the obstruction space $\mathscr{H}^1(\mathscr{E}^{\vee}|_x)$.

It is proved in [84, Proposition 2.8] that M is locally written as a critical locus. Indeed, let $U \subset N$ be an open subset such that the perfect obstruction theory (6.11) restricted to U is given as in the diagram (1.13), i.e.

$$\mathscr{E}|_U \cong (V^{\vee}|_U \xrightarrow{ds} \Omega_A|_U) \tag{6.12}$$

for a closed embedding $U \hookrightarrow A$ to a smooth A and a vector bundle $V \to A$ with a section s. Then there is the following function on the total space $V^{\vee} \to A$:

$$f : V^{\vee} \longrightarrow \mathbb{A}^1, \ (x, v) \longmapsto \langle s(x), v \rangle, \tag{6.13}$$

where $x \in A$ and $v \in V^{\vee}|_x$. Then one can show that $M \times_N U$ is isomorphic to the critical locus of f (see [84, Proposition 2.8]). One can then show that the above local data glue to give a canonical d-critical structure on M.

There is also canonical orientation data on M. Indeed, from (6.12), we have the isomorphism $K_{V^{\vee}}|_{\mathrm{Crit}(f)} \cong \pi|_U^* \det(\mathscr{E}|_U)$, so the virtual canonical line bundle of M is isomorphic to $\pi^* \det(\mathscr{E})^{\otimes 2}$. The canonical orientation data is given by $\pi^* \det(\mathscr{E})$. Therefore there is a canonical perverse sheaf ϕ_M on the dual obstruction cone $M \to N$, given as gluing of vanishing cycle sheaves of the functions (6.13).

Remark 6.7 From derived algebraic geometry, the dual obstruction cone is the classical truncation of a (-1)-shifted cotangent scheme. Namely let $\mathfrak{N}$ be a quasi-smooth derived scheme with classical truncation $N = t_0(\mathfrak{N})$. Then $\mathbb{L}_{\mathfrak{N}}|_N \to \tau_{\geq -1} \mathbb{L}_N$ is a perfect obstruction theory (see Remark 1.9). The (-1)-*shifted cotangent derived scheme* over $\mathfrak{N}$ is defined by

$$\Omega_{\mathfrak{N}}[-1] := \mathrm{Spec}_{\mathfrak{N}} \mathrm{Sym}^{\bullet}(\mathbb{T}_{\mathfrak{N}}[1]) \longrightarrow \mathfrak{N}.$$

Then the dual obstruction cone M is the classical truncation of $\Omega_{\mathfrak{N}}[-1]$. The (-1)-shifted cotangent scheme is (-1)-shifted symplectic (see [39, Theorem 2.4]), so its classical truncation M admits a canonical d-critical structure (see Remark 6.2). Under the above setting, it is proved by Kinjo [102] that there is an isomorphism $\pi_!\phi_M \cong \mathbb{Q}_N[\mathrm{vdim}N]$, where $\pi: M \to N$ is the projection and $\mathrm{vdim}N$ is the rank of $\mathbb{L}_{\mathfrak{M}}|_N$. The above isomorphism relates cohomological DT invariants on M with Borel–Moore homologies of N.

Remark 6.8 The dual obstruction cones appear in the study of moduli spaces of sheaves on local surfaces. Let S be a smooth projective surface and take the total space of its canonical line bundle (called *local surface*)

$$\pi: X = \mathrm{Tot}(K_S) \longrightarrow S. \tag{6.14}$$

The local surface X is a non-compact CY 3-fold.

Although X is non-compact, the stability of sheaves and their moduli spaces make sense for compactly supported sheaves on X. For an ample divisor H on S and $v \in H^*(S, \mathbb{Q})$, we denote by $M_X^{H\text{-st}}(v)$ the moduli space of π^*H-stable compactly supported coherent sheaves E on X such that $\mathrm{ch}(\pi_*E) = v$. If we assume that any stable sheaf $[E] \in M_X^{H\text{-st}}(v)$ pushes-forward to a H-stable sheaf π_*E on S, we have the morphism

$$\pi_*: M_X^{H\text{-st}}(v) \longrightarrow M_S^{H\text{-st}}(v), \ E \longmapsto \pi_*E. \tag{6.15}$$

Assuming that a universal sheaf $\mathcal{U} \in \mathrm{Coh}(S \times M_S^{H\text{-st}}(v))$ exists, we have the natural perfect obstruction theory on $M_S^{H\text{-st}}(v)$ (see Remark 1.8)

$$(\mathbf{R}p_{M*}\mathbf{R}\mathcal{H}om(\mathcal{U}, \mathcal{U})_0[1])^{\vee} \longrightarrow \tau_{\geq -1}\mathbb{L}_{M_S^{H\text{-st}}(v)}.$$

One can check that $M_X^{H\text{-st}}(v)$ is the dual obstruction cone over $M_S^{H\text{-st}}(v)$. Indeed, the fiber of the map (6.15) at $[F] \in M_S^{H*}(v)$ is given by the set of $\pi_*\mathcal{O}_X$-module structures on F. This is identified with the set of morphisms $F \to F \otimes K_S$, so the fiber of (6.15) is given by

$$\pi_*^{-1}([F]) = \mathrm{Hom}(F, F \otimes K_S) \cong \mathrm{Ext}^2(F, F)^{\vee},$$

which is dual to the obstruction space.

6.6　PBW Theorem for Cohomological DT Theory of Quivers with Super-Potentials

In this section, we explain the result of Davison–Meinhardt [52] on the *Poincaré–Birkhoff–Witt (PBW) theorem* for cohomological DT theory of quivers with super-potentials. Below we use the notation in Chap. 3.

Let $Q = (Q_0, Q_1, s, t)$ be a quiver. We say that Q is *symmetric* if the following condition holds for any $i, j \in Q_0$:

$$\{e \in Q_1 : s(e) = i, t(e) = j\} = \{e \in Q_1 : s(e) = j, t(e) = i\}.$$

For $v \in \Gamma_Q := \mathbb{Z}_{\geq 0}^{Q_0}$, let $\mathscr{M}_{(Q,W)}(v)$ be the moduli stack of (Q, W)-representations of dimension vector v. By (3.7), it is the critical locus of the function $\mathrm{tr}(W)$ on the smooth ambient stack $\mathscr{M}_Q(v)$. We have the associated perverse sheaf

$$\phi_{\mathrm{tr}(W)} \in \mathrm{Perv}(\mathscr{M}_{(Q,W)}(v)).$$

Its hyper-cohomology is the associated cohomological DT invariant. By taking the direct sum, we obtain the object in the (unbounded) derived category of sheaves of $\mathbb{Q}$-vector spaces over Γ_Q

$$\mathbb{H}^*(\mathscr{M}_{(Q,W)}, \phi_{\mathrm{tr}(W)}) := \bigoplus_{v \in \Gamma_Q} \mathbb{H}^*(\mathscr{M}_{(Q,W)}(v), \phi_{\mathrm{tr}(W)}) \in D(\Gamma_Q).$$

Here regard $\mathbb{H}(\mathscr{M}_{(Q,W)}(v), \phi_{\mathrm{tr}(W)})$ as a sheaf on Γ_Q supported at $v \in \Gamma_Q$.

On the other hand, recall that the function $\mathrm{tr}(W)$ on $\mathscr{M}_Q(v)$ factors through a function $\overline{\mathrm{tr}}(W)$ from the good moduli space $M_Q(v)$ as in the diagram (3.5). The BPS sheaf is defined as follows:

Definition 6.6 (*[52, Sect. 4.2]*) The *BPS sheaf* is a perverse sheaf on $M_Q(v)$ defined by

$$\mathscr{BPS}_{(Q,W)}(v) := \begin{cases} \Phi^p_{\overline{\mathrm{tr}}(W)}(\mathrm{IC}(M_Q(v))), & \text{if } M_Q^s(v) \neq \emptyset, \\ 0, & \text{if otherwise.} \end{cases}$$

Here $M_Q^s(v) \subset M_Q(v)$ is an open subset corresponding to simple Q-representations.

We also have the following object:

$$\mathbb{H}^*(M_Q, \mathscr{BPS}_{(Q,W)}) := \bigoplus_{v \in \Gamma_Q} \mathbb{H}^*(M_Q(v), BPS_{(Q,W)}) \in D(\Gamma_Q).$$

We prepare some more notation. By taking the sum, we have the map $+ \colon \Gamma_Q \times \cdots \times \Gamma_Q \to \Gamma_Q$. For $F \in D(\Gamma_Q)$, we define

$$\mathrm{Sym}^{\bullet}_+(F) := \bigoplus_{n \geq 0} +_*(\overbrace{F \boxtimes \cdots \boxtimes F}^{n})^{S_n} \in D(\Gamma_Q).$$

The following is one of main results by Davison–Meinhardt [52].

Theorem 6.3 ([52, Theorem A]) *There exists an isomorphism in* $D(\Gamma_Q)$

$$\mathbb{H}^*(\mathcal{M}_{(Q,W)}, \phi_{\mathrm{tr}(W)}) \cong \mathrm{Sym}^{\bullet}_+ \left(H_{\mathrm{vir}}(B\mathbb{C}^*) \otimes \mathbb{H}^*(M_Q, \mathscr{BPS}_{(Q,W)}) \right).$$

Here $H_{\mathrm{vir}}(B\mathbb{C}^*)$ *is given by*

$$H_{\mathrm{vir}}(B\mathbb{C}^*) := \bigoplus_{k \geq 0} \mathbb{Q}[-2k-1].$$

Remark 6.9 The perverse sheaf of vanishing cycles ϕ_f is upgraded to a monodromic mixed Hodge module, and the isomorphism in Theorem 6.3 can be formulated as an isomorphism in the derived category of monodromic mixed Hodge module on Γ_Q. There is also a relative version of the isomorphism in Theorem 6.3, formulated as an isomorphism in the derived category of monodromic mixed Hodge modules over M_Q. See [52, Theorem A] for details.

Remark 6.10 The result of Theorem 6.3 implies some integrality statement of generalized DT invariants on the quiver with super-potential (Q, W). Indeed, from Theorem 6.3, one can show that (though not explicitly written in the references)

$$\mathrm{DT}_{\theta=0}(v) = \sum_{k \geq 1, k | v} \frac{1}{k^2} \chi \left(\mathbb{H}^*(M_Q(v/k), \mathscr{BPS}_{(Q,W)}) \right).$$

Here the LHS is the generalized DT invariant for (Q, W) with trivial stability. Namely, a version of Conjecture 2.1 for a quiver with super-potential is true.

Chapter 7
Gopakumar–Vafa Invariants

Abstract In the last chapter, we explained a cohomological refinement of DT invariants via gluing of perverse sheaves of vanishing cycles. The resulting perverse sheaves on moduli spaces of one-dimensional stable sheaves are also used to define Gopakumar–Vafa invariants in [128]. In this chapter, we give an introduction to Gopakumar–Vafa invariants defined in [128] and discuss the conjecture relating GV invariants with PT invariants.

7.1 GW/PT/GV Correspondence Conjecture

Recall that in Conjecture 5.2, we discussed a conjectural relationship between rank one DT invariants and GW invariants. The former is equivalent to PT invariants by Theorem 5.2. However, neither of these invariants is an 'ideal' curve-counting invariant in the following sense:

(i) The GW invariants are rational numbers and not necessarily integers. For an ideal curve-counting, we want integer valued invariants.
(ii) The DT or PT invariants are integers, but they do not only count curves but also involve points in CY 3-folds.

From the perspective of string duality between type IIA and M theory, Gopakumar–Vafa [70] proposed the following conjecture:

Conjecture 7.1 ([70]) For a smooth projective CY 3-fold X, there exist integer-valued invariants

$$n_{g,\beta} \in \mathbb{Z}, \ g \geq 0, \ \beta \in H_2(X, \mathbb{Z}) \tag{7.1}$$

which vanish for sufficiently large g and which determine the Gromov–Witten series by the identity

$$\sum_{\beta > 0, g \geq 0} \mathrm{GW}_{g,\beta} \lambda^{2g-2} t^{\beta} = \sum_{\beta > 0, g \geq 0, k \geq 1} \frac{n_{g,\beta}}{k} \left(2 \sin \left(\frac{k\lambda}{2} \right) \right)^{2g-2} t^{k\beta}. \tag{7.2}$$

Y. Toda, *Recent Progress on the Donaldson–Thomas Theory*,
SpringerBriefs in Mathematical Physics 43,
https://doi.org/10.1007/978-981-16-7838-7_7

Fig. 7.1 GW/PT/GV
correspondence

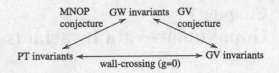

The invariants (7.1) are called *Gopakumar–Vafa (GV) invariants*. Under GW/DT/PT correspondence (Fig. 7.1), the invariants $n_{g,\beta}$ should also satisfy the identity in Conjecture 5.1. As we will see below (see (7.8)), if there is only a finite number of curves with homology class β, and each curve is smooth and super-rigid, then $n_{g,\beta}$ directly counts the number of such curves with genus g. In this sense, the invariant $n_{g,\beta}$ should be an ideal curve-counting invariant.

The original idea of Gopakumar–Vafa is to use $sl_2 \times sl_2$-action on cohomologies of moduli spaces of D2-branes, which are mathematically moduli spaces of one dimensional stable sheaves on CY 3-folds. Later Hosono–Saito–Takahashi [77] and Kiem–Li [100] proposed a mathematical definition of GV invariants following the approach by Gopakumar–Vafa [70]. However, it turns out that their definitions do not match with Conjecture 7.1 or Conjecture 5.1 (see Remark 7.8). In this chapter, we will explain the author's proposal with Maulik [128] on a mathematical definition of GV invariants via perverse sheaves of vanishing cycles, which modifies earlier approaches by Hosono–Saito–Takahashi [77] and Kiem–Li [100].

7.2 Moduli Spaces of One-Dimensional Stable Sheaves

Let X be a smooth projective CY 3-fold with a polarization H. For a coherent sheaf F on X with dim Supp$(F) = 1$, its *fundamental one cycle* $l(F)$ is defined by

$$l(F) := \sum_{\eta \in X, \dim \overline{\eta} = 1} \text{length}_\eta(F) \cdot \overline{\eta}. \tag{7.3}$$

For $\beta \in H_2(X, \mathbb{Z})$, we denote by $M_X(\beta)$ the moduli space of H-stable sheaves (see Definition 1.1) F on X with $([F], \chi(F)) = (\beta, 1)$, where $[F]$ is the homology class of the one cycle $l(F)$.

Remark 7.1 (1) The condition $([F], \chi(F)) = (\beta, n)$ is equivalent to

$$\text{ch}(F) = (0, 0, \beta, n) \in H^{2*}(X, \mathbb{Q}).$$

So $M_X(\beta)$ is the same as $M_X^{H\text{-st}}(0, 0, \beta, n)$ in the notation of Section 1.1.

(2) From the condition $\chi(F) = 1$, it is easy to check that F is H-stable if and only if it is H-semistable, if and only if for any subsheaf $0 \neq F' \subsetneq F'$ we have $\chi(F') \leq 0$. In particular, $M_X(\beta)$ is independent of H, which satisfies the ss=st condition. Moreover,

the $\chi(F) = 1$ condition obviously implies the condition (1.8), so $M_X(\beta)$ is a fine moduli space.

We also denote by $\mathrm{Chow}_X(\beta)$ the *Chow variety* parameterizing effective one cycles $Z \subset X$ whose homology class is β. Here we refer to [105, I.3] for details on the precise definition and the construction of the Chow variety.

Remark 7.2 Contrary to the moduli spaces of sheaves, the Chow variety $\mathrm{Chow}_X(\beta)$ is only constructed as a reduced scheme, as there is no natural deformation–obstruction theory on it.

We have the natural morphism, called *Hilbert–Chow morphism*

$$\pi_\beta \colon M_X^{\mathrm{red}}(\beta) \longrightarrow \mathrm{Chow}_X(\beta), \quad F \mapsto l(F). \tag{7.4}$$

For a smooth curve $C \subset X$ of genus g and homology class β, the fiber of π_β at the corresponding one cycle $[C] \in \mathrm{Chow}_X(\beta)$ is $\mathrm{Pic}_C(g)$, the moduli space of degree g line bundles on C. However, for a singular one cycle or non-reduced cycle, the fiber of π_β at such one cycles are in general intricate.

Example

Let $f \colon X \to S$ be an elliptically fibered CY 3-fold with a section and integral fibers. Let $[F]$ be a fiber class of f, and take $\beta = r[F]$. Then we have the commutative diagram (see [128, Sect. 9.2])

$$\begin{array}{ccc} X & \xrightarrow{\;\cong\;} & M_X(\beta) \\ {\scriptstyle \pi}\Big\downarrow & & \Big\downarrow{\scriptstyle \pi_\beta} \\ S & \lhook\joinrel\longrightarrow & \mathrm{Sym}^r(S). \end{array} \tag{7.5}$$

Here the right arrow is the Hilbert–Chow morphism, and the bottom arrow is the diagonal map. The top isomorphism is obtained as in [43, Lemma 2.1].

7.3 Gopakumar–Vafa Invariants via Vanishing Cycles

By Theorem 6.2, there exists a canonical d-critical structure s on $M_X(\beta)$. So given its orientation data, by Theorem 6.1 there exists a perverse sheaf $\phi_{M_X(\beta)}$ obtained as a gluing of locally defined perverse sheaves of vanishing cycles. In [128], we proposed a definition of GV invariants as follows:

Definition 1 ([128, Definition 3.7]) We define the GV invariants $n_{g,\beta}$ by the identity

$$\sum_{i\in\mathbb{Z}} \chi({}^{p}\mathcal{H}^{i}(\mathbf{R}\pi_{\beta*}\phi_{M_X(\beta)}))y^{i} = \sum_{g\geq 0} n_{g,\beta}(y^{\frac{1}{2}} + y^{-\frac{1}{2}})^{2g}. \tag{7.6}$$

Here ${}^{p}\mathcal{H}^{i}(-)$ is the i-th cohomology functor with respect to the perverse t-structure. By the self-duality of $\phi_{M_X(\beta)}$ and the Verdier duality, the LHS of (7.6) is uniquely written as the form of the RHS. The GV invariants in Definition 1 are obviously integers, which vanish for sufficiently large g.

The formulation of GV invariants as (7.6) depends on a choice of an orientation. We impose an additional restriction that the orientation data is trivial along the fibers of the Hilbert–Chow morphism. Let $K^{\text{vir}}_{M_X(\beta)}$ be the virtual canonical line bundle on $M_X(\beta)$, which is isomorphic to the determinant line bundle by (6.8).

Definition 7.1 Orientation data $\sqrt{K^{\text{vir}}_{M_X(\beta)}}$ of $M_X(\beta)$ is called *Calabi–Yau (CY)* if there is a line bundle $\mathscr{L}$ on $\text{Chow}_X(\beta)$ such that $\sqrt{K^{\text{vir}}_{M_X(\beta)}} \cong \pi_{\beta}^{*}\mathscr{L}$.

Once we take CY orientation data, the definition of GV invariants in (7.6) do not depend on a choice of CY orientation data (see [128, Lemma A.5]).

Remark 7.3 After our paper [128] appeared, Joyce–Upmeier [93] constructed canonical orientation data on $M_X(\beta)$ from gauge theory. It is not known whether their canonical orientation data is CY or not.

Remark 7.4 A necessary condition for the existence of CY orientation data is that $K^{\text{vir}}_{M_X(\beta)}$ is trivial along the fibers of the Hilbert–Chow morphism π_{β}. By the Riemann-Roch computation, it is easy to see that $K_{M_X(\beta)}$ is numerically trivial on fibers of π_{β}.

Remark 7.5 A phenomenon related to Remark 7.4 is that, in all the cases considered in [128], the d-critical locus $M_X(\beta)$ admits d-critical charts locally on the Chow variety, and moreover the smooth ambient space has a trivial canonical line bundle. More precisely, for any point $p \in \text{Chow}_X(\beta)$, there exists an open subset $p \in U \subset \text{Chow}_X(\beta)$ such that we have the commutative diagram

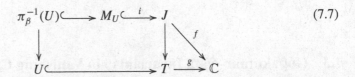

$$\tag{7.7}$$

such that (M_U, J, f, i) is a d-critical chart of $(M_U, s|_{M_U})$. Here each arrow $\hookrightarrow$ is a closed immersion, $M_U \subset M_X(\beta)$ is an open subscheme whose reduced part is $\pi_{\beta}^{-1}(M_U)$, and J is a smooth variety satisfying $K_J \cong \mathscr{O}_J$. In this case, in particular, we have CY orientation data $K_J|_{M_U} \cong \mathscr{O}_{M_U}$ on M_U. However, it is not known whether we have such d-critical charts (7.7) in general.

Remark 7.6 In $g = 0$, by substituting $y = -1$ the formula (7.6) immediately implies that (see the identity (6.9))

$$n_{0,\beta} = \chi(\mathbb{H}^*(M_X(\beta), \phi_{M_X(\beta)})) = N_{1,\beta}.$$

Here $N_{1,\beta}$ is the DT invariant of one dimensional sheaves in Theorem 5.3. The above identity is compatible with Katz's proposal of genus zero GV invariants [95].

Remark 7.7 Instead of $M_X(\beta) = M_X^{H\text{-st}}(0, 0, \beta, 1)$, one can also define GV invariants from the moduli spaces $M_X^{H\text{-ss}}(0, 0, \beta, n)$ (see [165, Definition 2.13]). The resulting invariant is independent of H (see [165, Theorem 5.5]), and conjectured to be the same invariant in Definition 1 (so it should be independent of n). The last conjecture is interpreted as a higher genus analogue of Conjecture 5.2. See [165, Conjecture 2.15] for details.

Example

Suppose that $\mathrm{Chow}_X(\beta)$ consists of finite number of points $[C_i]$ for smooth super-rigid curves $C_i \subset X$ with $1 \le i \le l$, i.e. $H^0(N_{C_i/X}) = 0$, with genus g_i. Then $M_X(\beta)$ is the disjoint union of $\mathrm{Pic}^{g_i}(C_i)$, so in particular, we have the orientation data which is trivial as a line bundle. The perverse sheaf $\phi_{M_X(\beta)}$ is isomorphic to $\mathbb{Q}_{\mathrm{Pic}_{C_i}(g_i)}[g_i]$ on $\mathrm{Pic}_{C_i}(g_i)$. Since $\mathrm{Pic}_{C_i}(g_i)$ is homeomorphic to $(S^1)^{2g_i}$, from the identity (7.6) together with

$$\sum_{i \in \mathbb{Z}} \dim \mathbb{H}^i((S^1)^{2g}, \mathbb{Q}[g]) y^i = (y^{\frac{1}{2}} + y^{-\frac{1}{2}})^{2g}$$

the GV invariants are calculated as

$$n_{g,\beta} = \sharp\{1 \le i \le l : g_i = g\}, \tag{7.8}$$

i.e. $n_{g,\beta}$ actually counts genus g-curves with homology class β.

Example

In the situation of the diagram (7.5), we have the decomposition

$$\mathbf{R}\pi_* \mathrm{IC}(X) = \mathrm{IC}(S)[1] \oplus V \oplus \mathrm{IC}(S)[-1],$$

where $V = R^1\pi_*\mathbb{Q}_X[2]$ is a perverse sheaf on S. For $s \in S$, let X_s be the fiber of π at s which is either an elliptic curve, rational curve with one node or a cusp. In any case, one can check that $R^1\pi_*\mathbb{Q}_X|_s = \mathbb{Q}^{2-e(X_s)}$. Then an easy calculation shows

$$\chi(\mathrm{IC}(S))y^{-1} + \chi(V) + \chi(\mathrm{IC}(S))y = -e(X) + e(S)(y^{\frac{1}{2}} + y^{-\frac{1}{2}})^2.$$

Therefore we obtain that $n_{0,\beta} = -e(X)$, $n_{1,\beta} = e(S)$ and $n_{g,\beta} = 0$ for $g \geq 2$.

In [128], we proposed that our definition of GV invariants agrees with GW and PT invariants.

Conjecture 7.2 ([128, Conjecture 3.13, 3.18])

(GW/GV conjecture) The GV invariants in Definition 1 satisfy the formula in Conjecture 7.1.

(PT/GV conjecture) The GV invariants in Definition 1 satisfy the formula in Conjecture 5.1.

If the above conjecture is true, then our invariants $n_{g,\beta}$ have to be deformation-invariant, which is not proved at this moment. In the following example, we see the deformation invariance of $n_{g,\beta}$ while the moduli spaces of stable sheaves and Chow varieties drastically change under deformations.

Example

Let S be an Enriques surface and E an elliptic curve. Let $\sigma : \widetilde{S} \to S$ be its K3 cover. We take a CY 3-fold

$$X = (\widetilde{S} \times E)/\langle\tau\rangle.$$

Here τ is an involution of $\widetilde{S} \times E$ which acts on $\widetilde{S}$ by the covering involution of σ and acts on E by $x \mapsto -x$. An Enriques surface S always admits an elliptic fibration $S \to \mathbb{P}^1$, with a double fiber $2C$. We take a curve class β in X by

$$\beta = ([C], 0) \in H_2(S, \mathbb{Z}) \oplus \mathbb{Z}[E] = H_2(X, \mathbb{Z}).$$

It is proved in [128, Proposition 7.6] that $n_{1,\beta} = 4$ and $n_{g,\beta} = 0$ for $g \neq 0$. Indeed, if S is a generic Enriques surface, C is a smooth elliptic curve and $\mathrm{Chow}_X(\beta)$ consists of four points $[C_i] \in \mathrm{Chow}_X(\beta)$ for super-rigid $C_i \cong C$. So the GV invariants are obtained as above from (7.8).

However, if S is not generic so that C is singular, then the moduli spaces drastically change. For example, if C is of type I_2, i.e. a cycle of two $\mathbb{P}^1$'s, the moduli space $M_X(\beta)$ together with the Hilbert–Chow morphism π_β in (7.4) is described in Fig. 7.3. Here each M_i, Γ_i is isomorphic to E, $\pi_\beta|_{M_i}$ restricts to the isomorphism $M_i \xrightarrow{\cong} \Gamma_i$, and each $C_i^{(j)}$ is isomorphic to $\mathbb{P}^1$ contracted by π_β.

Remark 7.8 In [77, 100], Hosono–Saito–Takahashi and Kiem–Li used $sl_2 \times sl_2$-action on some cohomology theories of $M_X(\beta)$ to define GV invariants. Our defini-

Fig. 7.2 Picture of HC morphism when C is of type I_2

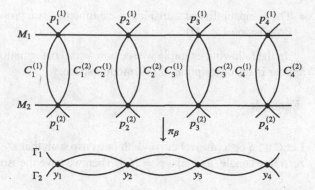

tion of GV invariants agrees with their definition when $M_X(\beta)$ is smooth. However, they are different when $M_X(\beta)$ is singular, for example their definitions in [77, 100] do not give a correct answer in the above example (see [128, Proposition 7.6]) (Fig. 7.2).

7.4 PT/GV Correspondence for Local Surfaces

A main result of [128] is to show PT/GV conjecture for local surfaces (see Remark 6.8). Although local surfaces are non-compact, the formulation of GV invariants make sense by considering compactly supported one dimensional stable sheaves on them. The PT invariants also make sense by considering compactly supported stable pair moduli spaces and the integrations of their Behrend functions. In the following, a curve class β is called *irreducible* if it is not written as a sum $\beta = \beta_1 + \beta_2$ for effective curve classes β_i.

Theorem 7.1 ([128, Theorem 5.15]) *Let S be a smooth projective surface satisfying* $H^1(\mathscr{O}_S) = 0$, *and* $X = \mathrm{Tot}(K_S)$ *the associated local surface. Then PT/GV conjecture in Conjecture 7.2 is true for any irreducible class* $\beta \in H_2(S, \mathbb{Z})$, *i.e. we have the following formula:*

$$\sum_{n \in \mathbb{Z}} P_{n,\beta} q^n = \sum_{g \geq 0} n_{g,\beta} (q^{\frac{1}{2}} + q^{-\frac{1}{2}})^{2g-2}.$$

The proof of the above theorem is complicated, and we refer to [128, Sect. 5] for details. The key point is to use the following:

- The description of the moduli space $M_X(\beta)$ on X in terms of dual obstruction cone over the moduli space of sheaves $M_S(\beta)$ on S (see Remark 6.8).
- The result of Maulik–Yun [129] and Migliorini–Shende [131] on generalized Macdonald formula for locally planar curves and their locally versal deformations.

- The compatibility of vanishing cycle functors with proper push-forwards (see [56, Proposition 4.2.11]).

In the following example we see how versal deformations of locally planar curves appear in the description of GV moduli space.

Example

Let $C \subset S$ be a rational curve with one cusp such that $\mathcal{O}_C(C)$ is a non-trivial degree zero line bundle, and take $\beta = [C]$. Then we have the isomorphism

$$C \xrightarrow{\cong} M_S(\beta), \; x \longmapsto I_x^{\vee},$$

where $I_x \subset \mathcal{O}_C$ is the ideal sheaf of x. Let $0 \in T \subset \mathbb{A}^2$ be a small open neighborhood and define $\pi_T \colon \mathscr{C} \to T$ by

$$\pi_T \colon \mathscr{C} = \{zy^2 = x^3 + t_1 xz^2 + t_2 z^3\} \subset \mathbb{P}^2 \times T \longrightarrow T,$$

where $[x : y : z]$ is the homogeneous coordinate of $\mathbb{P}^2$, (t_1, t_2) is the coordinate of T and the right arrow is the projection. The generic fiber of π_T is a smooth elliptic curve, and the π_T-relative compactified Jacobian is isomorphic to $\mathscr{C}$ itself. Since $M_X(\beta)$ is cut out by $t_1 = t_2 = 0$ in $\mathscr{C}$, the dual obstruction cone construction in Remark 6.8 yields the diagram

$$
\begin{array}{ccc}
M_X(\beta) \xrightarrow{\cong} \{df = 0\} \lhook\joinrel\longrightarrow & \mathscr{C} \times \mathbb{A}^2 & \\
\pi_\beta \downarrow \qquad\qquad & \pi_T \times \mathrm{id} \downarrow \quad \searrow f & \\
\mathrm{Chow}_X(\beta) \xrightarrow{\cong} \mathbb{A}^2 \overset{i}{\lhook\joinrel\longrightarrow} & T \times \mathbb{A}^2 \xrightarrow{g} & \mathbb{A}^1.
\end{array}
$$

Here the bottom arrows are given by $i(u_1, u_2) = (0, u_1, u_2)$ and $g(t_1, t_2, u_1, u_2) = t_1 u_1 + t_2 u_2$. Namely we have the Chow-local d-critical chart as in Remark 7.5.

Since $K_{\mathscr{C}} \cong \mathcal{O}_{\mathscr{C}}$, by taking CY orientation data we have $\phi_{M_X(\beta)} = \phi_f$. By the compatibility of vanishing cycle functors with proper push-forwards and perverse cohomologies, we have

$$
{}^p \mathcal{H}^i (\mathbf{R}\pi_{\beta *} \phi_{M_X(\beta)}) \cong {}^p \mathcal{H}^i (\Phi_g^p (\mathbf{R}(\pi_T \times \mathrm{id})_* \mathrm{IC}_{\mathscr{C} \times \mathbb{A}^2}))
$$

$$
\cong \Phi_g^p ({}^p \mathcal{H}^i (\mathbf{R}(\pi_T \times \mathrm{id})_* \mathrm{IC}_{\mathscr{C} \times \mathbb{A}^2})).
$$

Here the second isomorphism holds since Φ_g^p preserves the perverse t-structure. From the above isomorphisms, one can easily calculate that $n_{0,\beta} = -2$, $n_{1,\beta} = 1$ and $n_{\geq 2, \beta} = 0$ (see [128, Sect. 5.11]).

Chapter 8
Some Future Directions

Abstract In this final short chapter, we discuss some possible future directions of the study on DT invariants. The DT theory is now a huge subject, and I emphasize that the topics in this chapter are only a part of future directions chosen from the author's preference. The topics we discuss here are not entirely out of reach at this moment, but rather ongoing research subjects which are not yet mature. We expect great progress on these topics in the coming ten years or so.

8.1 Categorical Donaldson–Thomas Invariants

As we discussed in Chap. 6, the moduli space of stable sheaves M on a CY 3-fold X is locally written as a critical locus, and we can define the cohomological DT invariant by gluing locally defined perverse sheaves of vanishing cycles. There exist further categorifications of vanishing cycle sheaves, namely the dg-categories of matrix factorizations. So it is natural to try to glue these dg-categories to define a global dg-category $\mathscr{DT}(M)$, which we call the *DT category*.

For each d-critical chart (U, Y, f, i), the dg-category of matrix factorizations $\mathrm{MF}(Y, f)$ consists of data (see [142])

$$\mathscr{P}_0 \underset{\alpha_1}{\overset{\alpha_0}{\rightleftarrows}} \mathscr{P}_1, \quad \alpha_0 \circ \alpha_1 = \cdot f, \quad \alpha_1 \circ \alpha_0 = \cdot f,$$

where each $\mathscr{P}_i$ is a coherent sheaf on Y. By [59], it is related to the vanishing cycle sheaf by the isomorphism of $\mathbb{Z}/2$-graded vector spaces over $\mathbb{C}((u))$,

$$\mathrm{HP}_*(\mathrm{MF}(Y, w)) \cong H^*(U, \phi_f) \otimes_{\mathbb{Q}} \mathbb{C}((u)).$$

We expect the existence of a dg-category $\mathscr{DT}(M)$, to be something like

$$\mathscr{DT}(M) = \lim_{(U,Y,f,i)} \mathrm{MF}(Y, w) \otimes (\text{twisting}). \tag{8.1}$$

Y. Toda, *Recent Progress on the Donaldson—Thomas Theory*,
SpringerBriefs in Mathematical Physics 43,
https://doi.org/10.1007/978-981-16-7838-7_8

Here the limit should be taking in the ∞-category of $\mathbb{Z}/2$-periodic dg-categories.

A construction of the dg-category (8.1) for a general CY 3-fold is not yet available. In [166], the author proposed a definition of $\mathscr{D}\mathscr{T}(M)$ in the local surface case, i.e. $X = \mathrm{Tot}_S(\omega_S)$ for a smooth projective surface S. We then proposed several wall-crossing conjectures on the DT category, based on d-critical analogue of D/K conjecture by Bondal–Orlov [22] and Kawamata [98]. One of the conjectures is the existence of a fully-faithful functor

$$\mathscr{D}\mathscr{T}(P_n(X, \beta)) \longhookrightarrow \mathscr{D}\mathscr{T}(I_n(X, \beta)), \tag{8.2}$$

where $I_n(X, \beta)$, $P_n(X, \beta)$ are DT/PT moduli spaces which are discussed in Sect. 5.3. The above conjecture for local surfaces is proved in [166] when β is a reduced curve class. We also refer to [187] for related wall-crossing formulas for derived categories of PT moduli spaces.

It is an interesting and important research subject to construct the DT category (8.1) in general, and formulate categorical wall-crossing conjectures such as (8.2).

8.2 Donaldson–Thomas Invariants on Calabi–Yau 4-Folds

As we mentioned in Remark 1.14, now there is also a definition of DT invariants on CY 4-folds (called DT4 invariants) and there have been some recent efforts to prove (or conjecture) some analogy of several stories discussed in this monograph for DT4 invariants. For example, there is an analogy of the formula (1.24) for DT4 invariants in [41], a DT4 version of Theorem 3.1 in [48]. There is also a definition of Gopakumar–Vafa invariants for CY 4-folds by Klemm–Pandharipande [104], and their sheaf theoretic interpretation via DT4 invariants is discussed in [43–46].

Here we focus on the conjectural formula of stable pair invariants of CY 4-folds in [44]. Let X be a CY 4-fold, and take $\beta \in H_2(X, \mathbb{Z})$ and $n \in \mathbb{Z}$. Similarly to the case of CY 3-folds, there is a moduli space of PT stable pair moduli space $P_n(X, \beta)$. The construction of the DT4 virtual class in [23, 42, 139] gives

$$[P_n(X, \beta)]^{\mathrm{vir}} \in H_{2n}(P_n(X, \beta), \mathbb{Z}).$$

There are at least two differences from the CY 3-fold case. The first one is that the construction of the DT4 virtual class requires a choice of orientation data whose canonical choice is not known. The second one is that the virtual dimension of $P_n(X, \beta)$ is not necessarily zero, so in order to define an interesting invariant we need to choose some insertions.

Let us ignore the issue of orientations for a moment. For $\gamma \in H^4(X, \mathbb{Z})$, let $\tau_0(\gamma) \in H^2(P_n(X, \beta), \mathbb{Z})$ be the class of the locus incident to γ. In [44], we considered the primary PT invariants for CY 4-folds

$$P_{n,\beta}(\gamma) := \int_{[P_n(X,\beta)]^{\mathrm{vir}}} \tau_0(\gamma)^n \in \mathbb{Z}.$$

We then conjectured the following identity:

$$\sum_{n,\beta} \frac{1}{n!} P_{n,\beta}(\gamma) q^n t^\beta = \prod_\beta \exp(q t^\beta)^{n_{0,\beta}(\gamma)} \cdot M(t^\beta)^{n_{1,\beta}}. \tag{8.3}$$

Here $n_{0,\beta}(\gamma)$, $n_{1,\beta}$ are genus zero and one GV invariants on CY 4-folds [104] and $M(t)$ is the MacMahon function. We note that there are no GV invariants with genus bigger than or equal to two in the CY 4-fold case as the virtual dimensions of moduli spaces of such curves are negative. The formula (8.3) is regarded as a CY4 version of the formula in (5.1).

The formula (8.3) is derived by a heuristic argument: ideally there are only smooth rational curves which form a one-dimensional family, super-rigid elliptic curves, with no higher genus curves. The formula (8.3) is true in such an ideal situation. In some sense, the geometry behind the formula (8.3) is simpler than the CY 3-fold case in Conjecture 5.1 because of the absence of higher genus GV invariants. However, proving the formula (8.3) is in general much more difficult, even in some concrete examples. One of the difficulties is that we cannot apply the Behrend function technique in the CY 4-fold case. There is also an issue of orientation: at this moment, we can state that the formula (8.3) holds for *some* choices of orientations.

In [45], the formula (8.3) is interpreted as wall-crossing in the derived category. On the other hand, Joyce [85] shows the existence of vertex algebra actions on homologies of several moduli spaces of objects, and conjectures wall-crossing formulas of DT4 invariants in terms of vertex algebra actions. So proving such wall-crossing formulas would lead to great progress on DT4 invariants, proving formulas such as (8.3) (also see [6]).

8.3 Cohomological/categorical Hall Algebras

As we mentioned in Remark 6.4, there should be an algebra structure on cohomological DT invariants called COHA, which is interpreted as 'algebra of BPS states' in string theory. The construction of such an algebra is an important research subject, e.g. it would be used to have a geometric-representation-type statement on DT moduli spaces, and also useful to formulate wall-crossing formulas of cohomological DT invariants.

Roughly speaking, the COHA should be defined as follows. For a CY 3-fold X, let $\mathcal{M}_X$ be the moduli stack of coherent sheaves on X with Chern character v. Given an orientation of $\mathcal{M}_X$, there is a gluing of perverse sheaves of vanishing cycles $\phi_{\mathcal{M}} \in \mathrm{Perv}(\mathcal{M}_X)$. On the other hand, let $\mathcal{M}_X^{\mathrm{ex}}$ be the moduli stack of short exact sequences of coherent sheaves on X and $p_i : \mathcal{M}_X^{ex} \to \mathcal{M}_X$ be the evaluation morphisms (see Sect. 2.2). Then conjecturally there should exist a morphism

$$\mu : p_1^! \phi_{\mathcal{M}} \otimes p_2^! \phi_{\mathcal{M}} \longrightarrow p_3^! \phi_{\mathcal{M}} \text{ [some shift]} \qquad (8.4)$$

such that for $\alpha, \beta \in \mathbb{H}^*(\mathcal{M}_X, \phi_{\mathcal{M}})$, the product $\alpha * \beta$ may be defined as

$$\alpha * \beta = p_{3!} \circ \mu(p_1^!(\alpha) \otimes p_2^!(\beta)).$$

Moreover, in order to have an associative algebra structure by the above construction, the orientation data should satisfy some strong conditions (see [93, Sect. 4.3]).

The construction of a morphism (8.4) is a part of Joyce's conjecture on shifted Lagrangians for a (-1)-shifted symplectic derived stack (see [4, Conjecture 5.18], [91, Conjecture 1.1]). It states that for an oriented (-1)-shifted symplectic derived stack $\mathfrak{M}$ and an oriented Lagrangian $i : \mathfrak{L} \to \mathfrak{M}$, there is a morphism in the derived category of constructible sheaves

$$\mu_{\mathfrak{L}} : \mathbb{Q}_{\mathfrak{L}}[\text{vdim}\mathfrak{L}] \longrightarrow i^! \phi_{\mathfrak{M}}.$$

Proving such a conjecture is certainly an important problem, as it will also give a construction of DT4 virtual classes via perverse sheaves, and Fukaya-type categories for holomorphic symplectic manifolds (see [24, Remark 6.15]). The recent work by Kinjo [103] is along the line toward proving the above Joyce conjecture.

If the categorical DT invariants $\mathcal{DT}(M)$ are defined as (8.1), the COHA should be also extended to categorified COHA, which is a functor

$$\mathcal{DT}(M) \times \mathcal{DT}(M) \longrightarrow \mathcal{DT}(M)$$

satisfying some associativity conditions. In the local surface case, the above categorified COHA is constructed in [186], using Porta-Sala two dimensional categorified Hall algebras [152]. In the case of quivers with super-potentials, the categorified COHA is defined by Pădurairu [143, 144]. He also used it to prove a K-theoretic version of PBW theorem, whose cohomological version is stated in Theorem 6.3.

A construction of such categorified COHA would also be important in giving categorical geometric-representation-type statements, and also the categorical wall-crossing formula (e.g. describing the semiorthogonal complement of (8.2) which categorifies DT/PT correspondence in Theorem 5.2).

References

1. Alper, J.: Good moduli spaces for Artin stacks. Ann. Inst. Fourier (Grenoble) **63**(6), 2349–2402 (2013)
2. Alper, J., Hall, J., Rydh, D.: A Luna étale slice theorem for algebraic stacks. Ann. Math. (2) **191**(3), 675–738 (2020)
3. Alper, J., Halpern-Leistner, D., Heinloth, J.: Existence of moduli spaces for algebraic stacks. ArXiv:1812.01128
4. Amorim, L., Ben-Bassat, O.: Perversely categorified Lagrangian correspondences. Adv. Theor. Math. Phys. **21**(2), 289–381 (2017)
5. Arcara, D., Bertram, A.: Bridgeland-stable moduli spaces for K trivial surfaces. With an appendix by Max Lieblich. J. Eur. Math. Soc. **15**, 1–38 (2013)
6. Arkadij, B.: Wall-crossing for zero-dimensional sheaves and Hilbert schemes of points on Calabi–Yau 4-folds. ArXiv:2102.01056
7. Bayer, A., Macrì, E., Stellari, P.: The space of stability conditions on abelian threefolds, and on some Calabi-Yau threefolds. Invent. Math. **206**(3), 869–933 (2016)
8. Bayer, A., Macri, E., Toda, Y.: Bridgeland stability conditions on 3-folds I: Bogomolov-Gieseker type inequalities. J. Algebr. Geom. **23**, 117–163 (2014)
9. Beentjes, S.V., Calabrese, J., Rennemo, J.V.: A proof of the Donaldson-Thomas crepant resolution conjecture. ArXiv:1810.06581
10. Behrend, K.: Gromov-Witten invariants in algebraic geometry. Invent. Math. **127**, 601–617 (1997)
11. Behrend, K.: Donaldson-Thomas type invariants via microlocal geometry. Ann. Math. **170**, 1307–1338 (2009)
12. Behrend, K., Bryan, J.: Super-rigid Donaldson-Thomas invariants. Math. Res. Lett. **14**, 559–571 (2007)
13. Behrend, K., Bryan, J., Szendrői, B.: Motivic degree zero Donaldson-Thomas invariants. Invent. Math. **192**, 111–160 (2013)
14. Behrend, K., Fantechi, B.: The intrinsic normal cone. Invent. Math. **128**, 45–88 (1997)
15. Behrend, K., Fantechi, B.: Symmetric obstruction theories and Hilbert schemes of points on threefolds. Algebr. Number Theory **2**, 313–345 (2008)
16. Behrend, K., Ronagh, P.: The inertia operator on the motivic Hall algebra. Compos. Math. **155**(3), 528–598 (2019)
17. Beilinson, A., Bernstein, J., Deligne, P.: Faisceaux pervers. Anal. Topol. Singul. Spaces I, Asterisque **100**, 5–171 (1982)

SpringerBriefs in Mathematical Physics 43,
https://doi.org/10.1007/978-981-16-7838-7

18. Ben-Bassat, O., Brav, C., Bussi, V., Joyce, D.: A 'Darboux Theorem' for shifted symplectic structures on derived Artin stacks, with applications. Geom. Topol. **19**, 1287–1359 (2015)

19. Benson, D.J.: Representations and Cohomology I. Cambridge University Press (1991)

20. Bernardara, M., Macrì, E., Schmidt, B., Zhao, X.: Bridgeland stability conditions on Fano threefolds. Épijournal Géom. Algébrique **1**, Art. 2, 24 (2017)

21. Bogomolov, F.A.: Holomorphic tensors and vector bundles on projective manifolds. Izv. Akad. Nauk SSSR Ser. Mat. **42**, 1227–1287 (1978)

22. Bondal, A., Orlov, D.: Semiorthogonal decomposition for algebraic varieties. Arxiv:9506012

23. Borisov, D., Joyce, D.: Virtual fundamental classes for moduli spaces of sheaves on Calabi-Yau four-folds. Geom. Topol. **21**(6), 3231–3311 (2017)

24. Brav, C., Bussi, V., Dupont, D., Joyce, D., Szendrői, B.: Symmetries and stabilization for sheaves of vanishing cycles. With an appendix by Jörg Schürmann. J. Singul. **11**, 85–151 (2015)

25. Brav, C., Bussi, V., Joyce, D.: A Darboux theorem for derived schemes with shifted symplectic structure. J. Am. Math. Soc. **32**(2), 399–443 (2019)

26. Bridgeland, T.: Flops and derived categories. Invent. Math. **147**, 613–632 (2002)

27. Bridgeland, T.: Stability conditions on triangulated categories. Ann. Math. **166**, 317–345 (2007)

28. Bridgeland, T.: Stability conditions on $K3$ surfaces. Duke Math. J. **141**, 241–291 (2008)

29. Bridgeland, T.: Spaces of stability conditions. Proc. Sympos. Pure Math. **80**, 1–21 (2009). Algebraic Geometry, Seattle (2005)

30. Bridgeland, T.: An introduction to motivic Hall algebras. Adv. Math. **229**, 102–138 (2012)

31. Bridgeland, T.: Hall algebras and curve-counting invariants. J. Am. Math. Soc. **24**, 969–998 (2011)

32. Bridgeland, T., King, A., Reid, M.: The McKay correspondence as an equivalence of derived categories. J. Am. Math. Soc. **14**, 535–554 (2001)

33. Bryan, J., Cadman, C., Young, B.: The orbifold topological vertex. Adv. Math. **229**, 531–595 (2012)

34. Bryan, J., Katz, S., Leung, N.C.: Multiple covers and integrality conjecture for rational curves on Calabi-Yau threefolds. J. Algebr. Geom. **10**, 549–568 (2001)

35. Bryan, J., Oberdieck, G., Pandharipande, R., Yin, Q.: Curve counting on abelian surfaces and threefolds. Algebr. Geom. **5**(4), 398–463 (2018)

36. Bryan, J., Steinberg, D.: Curve counting invariants for crepant resolutions. Trans. Am. Math. Soc. **368**(3), 1583–1619 (2016)

37. Bussi, V., Joyce, D., Meinhardt, S.: On motivic vanishing cycles of critical loci. J. Algebr. Geom. **28**, 405–438 (2019)

38. Calabrese, J.: Donaldson-Thomas invariants and flops. J. Reine Angew. Math. **716**, 103–145 (2016)

39. Calaque, D.: Shifted cotangent stacks are shifted symplectic. Ann. Fac. Sci. Toulouse Math. (6) **28**(1), 67–90 (2019)

40. Căldăraru, A.: Nonfine moduli spaces of sheaves on $K3$ surfaces. Int. Math. Res. Not. **20**, 1027–1056 (2002)

41. Cao, Y., Kool, M.: Zero-dimensional Donaldson-Thomas invariants of Calabi-Yau 4-folds. Adv. Math. **338**, 601–648 (2018)

42. Cao, Y., Leung, N.C.: Donaldson-Thomas theory for Calabi-Yau 4-folds. ArXiv:1407.7659

43. Cao, Y., Maulik, D., Toda, Y.: Genus zero Gopakumar-Vafa type invariants for Calabi-Yau 4-folds. Adv. Math. **338**, 41–92 (2018)

44. Cao, Y., Maulik, D., Toda, Y.: Stable pairs and Gopakumar-Vafa type invariants for Calabi-Yau 4-folds, to appear. J. Eur. Math. Soc. ArXiv:1902.00003

45. Cao, Y., Toda, Y.: Curve counting via stable objects in the derived category of Calabi-Yau 4-folds. ArXiv:1909.04897

46. Cao, Y., Toda, Y.: Gopakumar-Vafa type invariants on Calabi-Yau 4-folds via descendent insertions. Commun. Math. Phys. **383**, 281–310 (2021)

47. Cao, Y., Toda, Y.: Tautological stable pair invariants of Calabi-Yau 4-folds. ArXiv:2009.03553

48. Cao, Y., Toda, Y.: Counting perverse coherent systems on Calabi-Yau 4-folds. ArXiv:2009.10909

49. Ciocan-Fontanine, I., Kapranov, M.: Virtual fundamental classes via dg-manifolds. Geom. Topol. **13**, 1779–1804 (2009)

50. Cox, D.A., Katz, S.: Mirror Symmetry and Algebraic Geometry, Mathematical Surveys and Monographs, vol. 68. American Mathematical Society (1999)

51. Davison, B.: The critical CoHA of a quiver with potential. Q. J. Math. **68**, 635–703 (2017)

52. Davison, B., Meinhardt, S.: Cohomological Donaldson-Thomas theory of a quiver with potential and quantum enveloping algebras. Invent. Math. **221**, 777–871 (2020)

53. Denef, F., Moore, G.: Split states, entropy enigmas, holes and halos. ArXiv:hep-th/0702146

54. Denef, J., Loeser, F.: Motivic exponential integrals and a motivic Thom-Sebastiani Theorem. Duke Math. J. **99**, 285–309 (1999)

55. Derksen, H., Weyman, J., Zelevinsky, A.: Quivers with potentials and their representations. Selecta Math. **14**, 59–119 (2008)

56. Dimca, A.: Sheaves in Topology. Springer, Berlin (2004)

57. Donovan, W., Wemyss, M.: Noncommutative deformations and flops. Duke Math. J. **165**(8), 1397–1474 (2016)

58. Douglas, M.: Dirichlet branes, homological mirror symmetry, and stability. In: Proceedings of the 2002 ICM, pp. 395–408 (2002)

59. Efimov, A.: Cyclic homology of categories of matrix factorizations. Int. Math. Res. Not. IMRN **12**, 3834–3869 (2018)

60. Feyzbakhsh, S., Thomas, R.P.: Curve counting and S-duality. ArXiv:2007.03037

61. Feyzbakhsh, S., Thomas, R.P.: Rank r DT theory from rank 0. ArXiv:2103.02915

62. Feyzbakhsh, S., Thomas, R.P.: Rank r DT theory from rank 1. ArXiv:2108.02828

63. Fukaya, K.: Deformation theory, homological algebra and mirror symmetry. In: Geometry and physics of branes (Como, 2001). Series High Energy Physics, Gravitation and Cosmology, pp. 121–209. IOP, Bristol (2003)

64. Fulton, W.: Intersection theory. Second edition, Ergebnisse der Mathematik und ihrer Grenzgebiete. 3. Folge, vol. 2. Springe (1998)

65. Gaitsgory, D., Rozenblyum, N.: A study in derived algebraic geometry. Vol. I. Correspondences and duality. Mathematical Surveys and Monographs, vol. 221. American Mathematical Society, Providence, RI (2017)

66. Gaitsgory, D., Rozenblyum, N.: A study in derived algebraic geometry. Vol. II. Deformations, Lie theory and formal geometry. Mathematical Surveys and Monographs, vol. 221. American Mathematical Society, Providence, RI (2017)

67. Gholampour, A., Kool, M.: Higher rank sheaves on threefolds and functional equations. Épijournal Géom. Algébrique **3**, Art. 17, 29 (2019)

68. Gieseker, D.: On a theorem of Bogomolov on Chern Classes of Stable Bundles. Am. J. Math. **101**, 77–85 (1979)

69. Ginzburg, V.: Calabi-Yau algebras. ArXiv:0612139

70. Gopakumar, R., Vafa, C.: M-theory and topological strings II. ArXiv:hep-th/9812127

71. Goresky, M., MacPherson, R.: Intersection homology. II. Invent. Math. **72**(1), 77–129 (1983)

72. Göttsche, L.: The Betti numbers of the Hilbert scheme of points on a smooth projective surface. Math. Ann. **286**, 193–207 (1990)

73. Graber, T., Pandharipande, R.: Localization of virtual classes. Invent. Math. **135**, 487–518 (1999)

74. Gulbrandsen, M.G.: Donaldson-Thomas invariants for complexes on abelian threefolds. Math. Z. **273**(1–2), 219–236 (2013)

75. Happel, D., Reiten, I., Smalø, S.O.: Tilting in abelian categories and quasitilted algebras. Mem. Am. Math. Soc. 120 (1996)

76. Hartshorne, R.: Residues and duality: Lecture notes of a seminar on the work of A. Grothendieck, given at Harvard 1963/1964. Lecture Notes in Mathematics (20). Springer, Berlin–New York (1966)

77. Hosono, S., Saito, M., Takahashi, A.: Relative Lefschetz actions and BPS state counting. Internat. Math. Res. Not. **15**, 783–816 (2001)
78. Hu, J., Li, W.P.: The Donaldson-Thomas invariants under blowups and flops. J. Differential. Geom. **90**, 391–411 (2012)
79. Hua, Z., Toda, Y.: Contraction algebra and invariants of singularities. Int. Math. Res. Not. IMRN **10**, 3173–3198 (2018)
80. Huang, M., Katz, S., Klemm, A.: Topological string on elliptic CY 3-folds and the ring of Jacobi forms. J. High Energy Phys. **125**, front matter+78 pp (2015)
81. Huybrechts, D.: Fourier-Mukai Transforms in Algebraic Geometry. Oxford Mathematical Monographs. The Clarendon Press, Oxford University Press (2006)
82. Huybrechts, D., Lehn, M.: Geometry of moduli spaces of sheaves. Aspects in Mathematics, vol. E31. Vieweg (1997)
83. Huybrechts, D., Thomas, R.P.: Deformation-obstruction theory for complexes via Atiyah-Kodaira-Spencer classes. Math. Ann. **346**, 545–569 (2010)
84. Jiang, Y., Thomas, R.P.: Virtual signed Euler characteristics. J. Algebr. Geom. **26**(2), 379–397 (2017)
85. Joyce, D.: Ringel–Hall style Lie algebra structures on the homology of moduli spaces. https://people.maths.ox.ac.uk/joyce/hall.pdf
86. Joyce, D.: Configurations in abelian categories I. Basic properties and moduli stack. Adv. Math. **203**, 194–255 (2006)
87. Joyce, D.: Configurations in abelian categories II. Ringel-Hall algebras. Adv. Math. **210**, 635–706 (2007)
88. Joyce, D.: Configurations in abelian categories III. Stability conditions and identities. Adv. Math. **215**, 153–219 (2007)
89. Joyce, D.: Configurations in abelian categories IV. Invariants and changing stability conditions. Adv. Math. **217**, 125–204 (2008)
90. Joyce, D.: A classical model for derived critical loci. J. Diff. Geom. **101**, 289–367 (2015)
91. Joyce, D., Safronov, P.: A Lagrangian neighbourhood theorem for shifted symplectic derived schemes. Ann. Fac. Sci. Toulouse Math. (6) **28**(5), 831–908 (2019)
92. Joyce, D., Song, Y.: A theory of generalized Donaldson-Thomas invariants. Mem. Am. Math. Soc. **217** (2012)
93. Joyce, D., Upmeier, M.: Orientation data for moduli spaces of coherent sheaves over Calabi-Yau 3-folds. Adv. Math. **381**, Paper No. 107627, 47 (2021)
94. Kapranov, M.: Noncommutative geometry based on commutator expansions. J. Reine Angew. Math. **505**, 73–118 (1998)
95. Katz, S.: Genus zero Gopakumar-Vafa invariants of contractible curves. J. Differ. Geom. **79**, 185–195 (2008)
96. Katz, S., Klemm, A., Vafa, C.: M-theory, topological strings and spinning black holes. Adv. Theor. Math. Phys. **3**, 1445–1537 (1999)
97. Kawai, T., Yoshioka, K.: String partition functions and infinite products. Adv. Theor. Math. Phys. **4**, 397–485 (2000)
98. Kawamata, Y.: D-equivalence and K-equivalence. J. Differ. Geom. **61**(1), 147–171 (2002)
99. Kawamata, Y.: On multi-pointed non-commutative deformations and Calabi-Yau threefolds. Compos. Math. **154**(9), 1815–1842 (2018)
100. Kiem, Y.H., Li, J.: Categorification of Donaldson-Thomas invariants via perverse sheaves. ArXiv:1212.6444
101. King, A.: Moduli of representations of finite-dimensional algebras. Quart. J. Math. Oxford Ser. **2**(45), 515–530 (1994)
102. Kinjo, T.: Dimensional reduction in cohomological Donaldson–Thomas theory. ArXiv:2102.01568
103. Kinjo, T.: Virtual classes via vanishing cycles. ArXiv:2109.06468
104. Klemm, A., Pandharipande, R.: Enumerative geometry of Calabi-Yau 4-folds. Commun. Math. Phys. **281**, 621–653 (2008)

105. Kollár, J.: Rational curves on algebraic varieties. Ergebnisse Math. Grenzgeb. (3) vol. 32. Springer (1996)
106. Kollár, J., Mori, S.: Birational geometry of algebraic varieties. Cambridge Tracts in Mathematics, vol. 134. Cambridge University Press, Cambridge (1998). With the collaboration of C. H. Clemens and A. Corti. Translated from the 1998 Japanese original
107. Kontsevich, M.: Enumeration of rational curves via torus actions. The moduli space of curves. Progr. Math. **129**, 335–368 (1995)
108. Kontsevich, M., Soibelman, Y.: Stability structures, motivic Donaldson-Thomas invariants and cluster transformations. ArXiv:0811.2435
109. Kontsevich, M., Soibelman, Y.: Cohomological Hall algebra, exponential Hodge structures and motivic Donaldson-Thomas invariants. Commun. Number Theory Phys. **5**(2), 231–352 (2011)
110. Koseki, N.: Stability conditions on threefolds with nef tangent bundles. Adv. Math. **372**, 107316, 29 (2020)
111. Laudal, O.: Noncommutative deformations of modules. Homol. Homotopy Appl. **4**, 357–396 (2002)
112. Laumon, G., Moret-Bailly, L.: Champs algébriques, Ergebnisse der Mathematik und ihrer Grenzgebiete, vol. 39. Springer, Berlin (2000)
113. Levine, M., Pandharipande, R.: Algebraic cobordism revisited. Invent. Math. **176**, 63–130 (2009)
114. Li, C.: On stability conditions for the quintic threefold. Invent. Math. **218**(1), 301–340 (2019)
115. Li, C.: Stability conditions on Fano threefolds of Picard number 1. J. Eur. Math. Soc. (JEMS) **21**(3), 709–726 (2019)
116. Li, J.: A degeneration formula of GW-invariants. J. Differ. Geom. **60**(2), 199–293 (2002)
117. Li, J.: Zero dimensional Donaldson-Thomas invariants of threefolds. Geom. Topol. **10**, 2117–2171 (2006)
118. Li, J., Tian, G.: Virtual moduli cycles and Gromov-Witten invariants of algebraic varieties. J. Am. Math. Soc. **11**, 119–174 (1998)
119. Li, J., Wu, B.: Good degeneration of Quot-schemes and coherent systems. Commun. Anal. Geom. **23**(4), 841–921 (2015)
120. Lieblich, M.: Moduli of complexes on a proper morphism. J. Algebr. Geom. **15**, 175–206 (2006)
121. Liu, Y.: Stability conditions on product varieties. J. Reine Angew. Math. **770**, 135–157 (2021)
122. Lo, J., Qin, Z.: Mini-walls for Bridgeland stability conditions on the derived category of sheaves over surfaces. Asian J. Math. **18**(2), 321–344 (2014)
123. Maciocia, A., Piyaratne, D.: Fourier-Mukai transforms and Bridgeland stability conditions on abelian threefolds. Algebr. Geom. **2**(3), 270–297 (2015)
124. Manschot, J.: Sheaves on $\mathbb{P}^2$ and generalized Appell functions. Adv. Theor. Math. Phys. **21**(3), 655–681 (2017)
125. Massey, D.B.: The Sebastiani-Thom isomorphism in the derived category. Compositio Math. **125**(3), 353–362 (2001)
126. Maulik, D., Nekrasov, N., Okounkov, A., Pandharipande, R.: Gromov-Witten theory and Donaldson-Thomas theory. I. Compositio. Math. **142**, 1263–1285 (2006)
127. Maulik, D., Thomas, R.P.: Sheaf counting on local K3 surfaces. Pure Appl. Math. Quart. **2018**, 419–441 (2018)
128. Maulik, D., Toda, Y.: Gopakumar-Vafa invariants via vanishing cycles. Invent. Math. **213**(3), 1017–1097 (2018)
129. Maulik, D., Yun, Z.: Macdonald formula for curves with planar singularities. J. Reine Angew. Math. **694**, 27–48 (2014)
130. Meinhardt, S.: An Introduction to (Motivic) Donaldson-Thomas Theory. Confluentes Mathematici **9**, 101–158 (2017)
131. Migliorini, L., Shende, V.: A support theorem for Hilbert schemes of planar curves. J. Eur. Math. Soc. **15**, 2353–2367 (2013)

132. Mochizuki, T.: Donaldson type invariants for algebraic surfaces. Lecture Notes in Mathematics, vol. 1972. Springer, Berlin (2009)
133. Morrison, A., Mozgovoy, S., Nagao, K., Szendrői, B.: Motivic Donaldson-Thomas invariants of the conifold and the refined topological vertex. Adv. Math. **230**, 2065–2093 (2012)
134. Nagao, K.: Donaldson-Thomas theory and cluster algebras. Duke Math. J. **162**(7), 1313–1367 (2013)
135. Nagao, K., Nakajima, H.: Counting invariant of perverse coherent sheaves and its wall-crossing. Int. Math. Res. Not. IMRN **17**, 3885–3938 (2011)
136. Nekrasov, N., Okounkov, A.: Membranes and sheaves. Algebr. Geom. **3**(3), 320–369 (2016)
137. Oberdieck, G., Piyaratne, D., Toda, Y.: Donaldson-Thomas invariants on abelian threefolds and Bridgeland stability conditions. ArXiv:1808.02735
138. Oberdieck, G., Shen, J.: Curve counting on elliptic Calabi-Yau threefolds via derived categories. J. Eur. Math. Soc. (JEMS) **22**(3), 967–1002 (2020)
139. Oh, J., Thomas, R.P.: Counting sheaves on Calabi-Yau 4-folds I. ArXiv:2009.05542
140. Okounkov, A.: Takagi lectures on Donaldson-Thomas theory. Jpn. J. Math. **14**, 67–133 (2019)
141. Olsson, M.: Algebraic Spaces and Stacks, American Mathematical Society Colloquium Publications, vol. 62. American Mathematical Society, Providence, RI (2016)
142. Orlov, D.: Matrix factorizations for nonaffine LG-models. Math. Ann. **353**(1), 95–108 (2012)
143. Pădurairu, T.: K-theoretic Hall algebras for quivers with potential. ArXiv:1911.05526
144. Pădurairu, T.: Categorical and K-theoretic Hall algebras for quivers with potential. ArXiv:2107.13642
145. Pandharipande, R., Pixton, A.: Gromov-Witten/Pairs correspondence for the quintic 3-fold. J. Am. Math. Soc. **30**(2), 389–449 (2017)
146. Pandharipande, R., Thomas, R.P.: Curve counting via stable pairs in the derived category. Invent. Math. **178**, 407–447 (2009)
147. Pandharipande, R., Thomas, R.P.: 13/2 ways of counting curves. Moduli Spaces London Math. Soc. Lect. Note Ser. **411**, 282–333 (2014)
148. Pandharipande, R., Thomas, R.P.: The Katz-Klemm-Vafa conjecture for K3 surfaces. Forum Math. **2016**, 1–111 (2016)
149. Pantev, T., Toën, B., Vaquie, M., Vezzosi, G.: Shifted symplectic structures. Publ. Math. IHES **117**, 271–328 (2013)
150. Piyaratne, D., Toda, Y.: Moduli of Bridgeland semistable objects on 3-folds and Donaldson-Thomas invariants. J. Reine Angew. Math. **747**, 175–219 (2019)
151. Polishchuk, A., Tu, J.: DG-resolutions of NC-smooth thickenings and NC-Fourier-Mukai transforms. Math. Ann. **360**, 79–156 (2014)
152. Porta, M., Sala, F.: Two dimensional categorified Hall algebras. ArXiv:1903.07253
153. Ringel, C.M.: Hall algebras and quantum groups. Invent. Math. **101**(3), 583–591 (1990)
154. Schmidt, B.: Counterexample to the generalized Bogomolov-Gieseker inequality for three-folds. Int. Math. Res. Not. IMRN **8**, 2562–2566 (2017)
155. Segal, E.: The A_∞ deformation theory of a point and the derived categories of local Calabi-Yaus. J. Algebr. **320**, 3232–3268 (2008)
156. Stanley, R.: Enumerative Combinatorics. Cambridge University Press (1999)
157. Stoppa, J.: D0–D6 states counting and GW invariants. Lett. Math. Phys. **102**, 149–180 (2012)
158. Szendrői, B.: Non-commutative Donaldson-Thomas theory and the conifold. Geom. Topol. **12**, 1171–1202 (2008)
159. Szendrői, B.: Cohomological Donaldson-Thomas theory. String-Math 2014. Proc. Sympos. Pure Math. **93**, 363–396. Am. Math. Soc., Providence, RI (2016)
160. Tanaka, Y., Thomas, R.P.: Vafa-Witten invariants for projective surfaces II: semistable case. Pure Appl. Math. Q. **13**(3), 517–562 (2017)
161. Tanaka, Y., Thomas, R.P.: Vafa-Witten invariants for projective surfaces I: stable case. J. Algebr. Geom. **29**, 603–668 (2020)
162. Taubes, C.H.: Casson's invariant and gauge theory. J. Differ. Geom. **31**(2), 547–599 (1990)
163. Thomas, R.P.: A holomorphic Casson invariant for Calabi-Yau 3-folds and bundles on $K3$-fibrations. J. Differ. Geom **54**, 367–438 (2000)

164. Thomas, R.P.: An obstructed bundle on a Calabi-Yau 3-fold. Adv. Theor. Math. Phys. **3**, 567–576 (2000)
165. Toda, Y.: Gopakumar-Vafa invariants and wall-crossing. ArXiv:1710.01843
166. Toda, Y.: Categorical Donaldson-Thomas theory on local surfaces. ArXiv:1907.09076
167. Toda, Y.: Moduli stacks and invariants of semistable objects on K3 surfaces. Adv. Math. **217**, 2736–2781 (2008)
168. Toda, Y.: Limit stable objects on Calabi-Yau 3-folds. Duke Math. J. **149**, 157–208 (2009)
169. Toda, Y.: Curve counting theories via stable objects I: DT/PT correspondence. J. Am. Math. Soc. **23**, 1119–1157 (2010)
170. Toda, Y.: Generating functions of stable pair invariants via wall-crossings in derived categories. Adv. Stud. Pure Math. **59**, 389–434 (2010). New developments in algebraic geometry, integrable systems and mirror symmetry (RIMS, Kyoto, 2008)
171. Toda, Y.: On a computation of rank two Donaldson-Thomas invariants. Commun. Number Theory Phys. **4**, 49–102 (2010)
172. Toda, Y.: Stability conditions and curve counting invariants on Calabi-Yau 3-folds. Kyoto J. Math. **52**, 1–50 (2012)
173. Toda, Y.: Stable pairs on local K3 surfaces. J. Differ. Geom. **92**, 285–370 (2012)
174. Toda, Y.: Bogomolov-Gieseker type inequality and counting invariants. J. Topol. **6**, 217–250 (2013)
175. Toda, Y.: Curve counting theories via stable objects II. DT/ncDT flop formula. J. Reine Angew. Math. **675**, 1–51 (2013)
176. Toda, Y.: Stability conditions and birational geometry of projective surfaces. Compos. Math. **150**, 1755–1788 (2014)
177. Toda, Y.: Flops and the S-duality conjecture. Duke Math. J. **164**, 2293–2339 (2015)
178. Toda, Y.: Non-commutative width and Gopakumar-Vafa invariants. Manuscripta Math. **148**, 521–533 (2015)
179. Toda, Y.: Stable pair invariants on Calabi-Yau 3-folds containing $\mathbb{P}^2$. Geom. Topol. **20**, 555–611 (2016)
180. Toda, Y.: Generalized Donaldson-Thomas invariants on the local projective plane. J. Differ. Geom. **106**(2), 341–369 (2017)
181. Toda, Y.: Gepner point and strong Bogomolov-Gieseker inequality for quintic 3-folds. Higher dimensional algebraic geometry—in honour of Professor Yujiro Kawamata's sixtieth birthday. Adv. Stud. Pure Math. **74**, 381–405. Math. Soc. Japan, Tokyo (2017)
182. Toda, Y.: Non-commutative thickening of moduli spaces of stable sheaves. Compos. Math. **153**(6), 1153–1195 (2017)
183. Toda, Y.: Moduli stacks of semistable sheaves and representations of Ext-quivers. Geom. Topol. **22**(5), 3083–3144 (2018)
184. Toda, Y.: Non-commutative deformations and Donaldson-Thomas invariants. Algebraic geometry: Salt Lake City 2015. Proc. Sympos. Pure Math., vol. 97, 611–631. Am. Math. Soc., Providence, RI (2018)
185. Toda, Y.: Hall algebras in the derived category and higher-rank DT invariants. Algebr. Geom. **7**(3), 240–262 (2020)
186. Toda, Y.: Hall-type algebras for categorical Donaldson-Thomas theories on local surfaces. Selecta Math. (N.S.) **26**(4), 64 (2020)
187. Toda, Y.: Semiorthogonal decompositions of stable pair moduli spaces via d-critical flips. J. Eur. Math. Soc. (JEMS) **23**(5), 1675–1725 (2021)
188. Moduli of objects in dg-categories: Toën, B., Vaquié, M. Ann. Sci. École Norm **40**, 387–444 (2007)
189. Toën, B.: Derived algebraic geometry. EMS Surv. Math. Sci. **1**(2), 153–240 (2014)
190. Vafa, C., Witten, E.: A strong coupling test of S-duality. Nucl. Phys. B **431** (1994)
191. Van den Bergh, M.: Three-dimensional flops and noncommutative rings. Duke Math. J. **122**(3), 423–455 (2004)
192. Young, B.: Computing a pyramid partition generating function with dimer shuffling. J. Combin. Theory Ser. A **116**, 334–350 (2009)

Index

© The Author(s), under exclusive license to Springer Nature Singapore Pte Ltd. 2021   103
Y. Toda, *Recent Progress on the Donaldson—Thomas Theory*,
SpringerBriefs in Mathematical Physics 43,
https://doi.org/10.1007/978-981-16-7838-7

Printed in the United States
by Baker & Taylor Publisher Services